Journey from Natural Numbers to Complex Numbers

Advances in Mathematics and Engineering

Series Editor: Mangey Ram, Department of Mathematics, Graphic Era University, Dehradun, Uttarakhand, India

The main aim of this focus book series is to publish the original articles that bring up the latest development and research in mathematics and its applications. The books in this series are in short form, ranging between 20,000 and 50,000 words or 100 to 125 printed pages, and encompass a comprehensive range of mathematics and engineering areas. It will include, but won't be limited to, mathematical engineering sciences, engineering and technology, physical sciences, numerical and computational sciences, space sciences, and meteorology. The books within the series line will provide professionals, researchers, educators, and advanced students in the field with an invaluable reference into the latest research and developments.

Recent Advancements in Software Reliability Assurance

Edited by Adarsh Anand and Mangey Ram

Linear Transformation

Examples and Solutions

Nita H. Shah and Urmila B. Chaudhari

Scientific Methods Used in Research and Writing

Mangey Ram, Om Prakash Nautiyal, and Durgesh Pant

Journey from Natural Numbers to Complex Numbers

Nita Shah and Vishnuprasad D. Thakkar

Partial Differential Equations: An Introduction

Nita Shah and Mrudul Y. Jani

For more information about this series, please visit: https://www.crcpress.com/Advances-in-Mathematics-and-Engineering/book-series/AME

Journey from Natural Numbers to Complex Numbers

Nita H. Shah and Vishnuprasad D. Thakkar

CRC Press
Taylor & Francis Group
Boca Raton London New York

CRC Press is an imprint of the
Taylor & Francis Group, an **informa** business

First edition published 2021
by CRC Press
6000 Broken Sound Parkway NW, Suite 300, Boca Raton, FL 33487-2742
and by CRC Press
2 Park Square, Milton Park, Abingdon, Oxon, OX14 4RN

© 2021 Taylor & Francis Group, LLC

CRC Press is an imprint of Taylor & Francis Group, LLC

Library of Congress Cataloging-in-Publication Data
Names: Shah, Nita H., author. I Vishnuprasad, Thakkar D., author. Title: Journey from natural numbers to complex numbers / Nita H. Shah and Thakkar D. Vishnuprasad. Description: Boca Raton : CRC Press, 2021. I Series: Advances in mathematics and engineering I Includes bibliographical references and index. Identifiers: LCCN 2020035251 (print) I LCCN 2020035252 (ebook) I ISBN 9780367613327 (hardback) I ISBN 9781003105244 (ebook) Subjects: LCSH: Numbers, Natural. I Numbers, Complex. Classification: LCC QA141 .S48 2021 (print) I LCC QA141 (ebook) I DDC 512.7--dc23 LC record available at https://lccn.loc.gov/2020035251LC ebook record available at https://lccn.loc.gov/2020035252

ISBN: 9780367613327 (hbk)
ISBN: 9781003105244 (ebk)

Typeset in Times
by Deanta Global Publishing Services, Chennai, India

Visit the www.routledge.com/ 9780367613327

Contents

Preface   vii
Author biographies   xiii

1 Natural Numbers   **1**
  1.1 Prerequisites   1
       1.1.1 Set Theory   1
       1.1.2 Relation   3
       1.1.3 Function   6
       1.1.4 Cardinality   8
       1.1.5 Algebra   10
  1.2 Positive Integers   14
       1.2.1 Positive Integers in Real Life   14
       1.2.2 Set Theoretic Definition of Natural Numbers   16
       1.2.3 Peano Axioms   17
       1.2.4 Ordering in Natural Numbers   19
       1.2.5 First Principle of Mathematical Induction   20
       1.2.6 Second Principle of Mathematical Induction   20
       1.2.7 Well-Ordering Principle   21
       1.2.8 Limitations of Natural Numbers   21
       1.2.9 Representation of Natural Numbers   21
              1.2.9.1 Hexadecimal System   23
       1.2.10 Number System Used by Computers   26
  1.3 Summary   29

2 Integers   **31**
  2.1 Informal Introduction of Integers   31
  2.2 Integers as Relation in Ordered Pairs of Natural Numbers   35
  2.3 Ordering in Ordered Pairs   36
  2.4 Operations in Ordered Pairs of Natural Numbers   36
  2.5 Properties of Binary Operations   37
  2.6 Interpretation of Relation and Operations   39
  2.7 Mapping of Ordered Pairs as Extension of Natural
       Numbers   42
  2.8 Representation of Integers   43
  2.9 Summary   45

3 Rational Numbers   **47**
  3.1   Informal Introduction of Rational Numbers   47
  3.2   Rational Numbers as Relation in Ordered Pairs
      of Integers   49
  3.3   Ordering in Ordered Pairs   50
  3.4   Operations in Ordered Pairs   50
  3.5   Properties of Binary Operations   50
  3.6   Interpretation of Relation and Operations   51
  3.7   Mapping of Ordered Pairs as Extension of Integers   52
  3.8   Representation of Rational Numbers   53
  3.9   Limitations of Rational Numbers   58
  3.10 Summary   59

4 Real Numbers   **61**
  4.1   Least Upper Bound Property   62
  4.2   Rational Cuts   62
  4.3   Dedekind Cuts   63
  4.4   Ordering in Cuts   64
  4.5   Binary Operations in Cuts   65
  4.6   Least Upper Bound Property   66
  4.7   Set of Cuts as Extension of Rational Numbers   66
  4.8   Cardinality of Set of Real Numbers   67
  4.9   Limitations of Real Numbers   69
  4.10 Summary   69

5 Complex Numbers   **71**
  5.1   Complex Numbers as Ordered Pairs of Real Numbers   71
  5.2   Binary Operations in Complex Numbers   71
  5.3   Introduction of Imaginary Numbers   72
  5.4   Representation of Complex Numbers   73
  5.5   Ordering in Complex numbers   76
  5.6   Cardinality of the Set of Complex Numbers   76
  5.7   Algebraic Numbers   77
  5.8   Summary   77

Index   79

Preface

This book is about different types of numbers. Counting process starts with positive whole numbers, known as natural numbers. The system is extended to have the negative whole numbers and zero in it. It is further extended to rational numbers, followed by real numbers, and ultimately by complex numbers. For natural numbers, integers, and rational numbers, before going for the formal study, we have started with real-life problems and introduced the basis of formal definition with real-life problems.

The formal study of the system has references to some topics in mathematics. For the sake of completeness, we have included a brief introduction of these topics in Chapter 1 [1–3]. We have started with set theory. Set theory starts with an informal definition of set, followed by the definition of subset, superset, null set, and universal set. Operations like union, intersection, and product are introduced. The collection of sets and union and intersection operations on the collection of sets are introduced. Power set is introduced, followed by the relation between two sets. A special case of the relation of sets with itself is taken. Properties of relations such as reflexivity, symmetry, transitivity, and antisymmetry are introduced. Based on these properties, equivalence relation and partial order are defined. Partition of the set on the basis of the equivalence relation is defined. The partial order is extended to the total order. On the basis of relation, functions are defined. Properties of functions and composite functions are defined. The concept of invertible functions is introduced. The cardinality of the set is defined using functions. Some interesting facts related to cardinality are proved or stated. Binary operations are introduced as a function from product set of a set with itself to the same set. Properties of binary operations such as commutativity and associativity are defined. Identity element and the inverse of the member operand are defined. The structure of a group is defined, and Abelian groups are defined. A set with two operations (one called addition and another multiplication) is introduced, and the distributivity of multiplication over addition is defined. Ring and fields are defined for the set with operations of addition and multiplication. This brief introduction of concepts can be used as a quick reference for all formal mathematics courses. With this background knowledge, we proceed to the number systems.

We have given an informal introduction of natural numbers. Addition and multiplication are explained with real-life physical activities and structures, keeping the formal definition of the operations in mind. Informal proof of commutativity and associativity is given. After the informal introduction of natural numbers, we move to the formal definition. Axiomatic definition of natural numbers is done using set-theoretic definition and Peano axioms. The formal definitions of addition and multiplication are given. Ordering is defined in natural numbers. The first principle and the second principle of mathematical induction and well-ordering principles are stated. The equivalence of three principles is stated. Prime numbers and composite numbers are defined. A major limitation of the absence of additive inverse and multiplicative inverse is identified, which forms the basis for the extension of natural numbers to integers. Coming back to the physical aspect of natural numbers, the representation of natural numbers is taken up. Position-based value system for writing numbers is explained by two models. This is generalized to any positive integer base greater than 1. Conversions of number from one base to another base are explained. Popular bases – 2 (binary), 8 (octal), and 16 (hexadecimal) – are taken up. Special cases of converting numbers to base, where the target base is the power of the source base or the source base is the power of the target base, are taken up. Computers use the binary system. The relation between binary system–based arithmetic and Boolean algebra is explained. The rationale of the usage of the octal and hexadecimal systems by computer professionals is explained. Binary Coded Decimal (BCD), another important system, is explained. In the next chapter, building up of integers using natural numbers is done.

We have used the trading activity to understand the extension of natural numbers to integers. We have a limited subtraction capability in natural numbers. We can subtract a smaller number from a larger number in natural numbers. Using this capability, there is an associated profit or loss, depending on which the price is larger. Two trades are considered equivalent if they give the same profit or same loss. If both the prices are equal, then the trade has a no-profit, no-loss. The addition of trades is done by combining trades, which is done by adding corresponding prices (purchase price and sales price). We have established that no-profit, no-loss trades are zero trades, and adding any of them to any trade does not change the trade outcome. We can equate profit trades to natural numbers, loss trades to corresponding negative trades, and no-profit, no-loss trades to zero trades. The repetition of trade and the reversal of trade are introduced for multiplication. The reversal of trades is considered as a negative repetition. Using this model, we have created the foundation for the formal method of extending natural numbers to integers. We define an equivalence relation on ordered pairs of natural numbers like an ordered pair of sales price and purchase price. We define equivalence relation for the ordered pairs similar to equal profit or loss trades. Similarly, we have defined addition and multiplication operations on ordered pairs [3]. It is established that operations on the ordered pairs maintain equivalence relation. The equivalence relation gives partition of ordered pairs. We have plotted all ordered pairs on XY plane. There is a unique line with 45 degree slope for each equivalent class in the plot. These lines intersect X-axis at some point. Intersection points are mapped to natural numbers for the classes (lines) the intersection point corresponds to the natural number on X-axis. Operations and ordering are consistent between classes and natural numbers. Hence, we can consider that the classes are the extension of natural numbers. They are interpreted for negative numbers of natural numbers, and arithmetic operations are interpreted in accordance with the definition. We have established that the set of integers with the definition of addition and multiplication is a commutative ring. The binary system introduced for the natural number is extended for negative integers. Mechanical counter–based model is used to explain the representation of negative numbers. The design philosophy of 2's complement for binary and 10's complement is explained. One major shortcoming of integers is the absence of multiplicative inverse. As a result, it is not always possible to divide integer by another integer to get integer result. We move to rational numbers in the next chapter.

We have used pizza distribution among a group of persons as an example to understand division. We start with cases where the number of pizzas is the multiple of the number of persons. Division process is considered as the inverse of multiplication. As another option, divisor is repeated subtracted and the count of subtraction iterations is the quotient of division. However, we would like to find the quotient when the dividend is not the multiple of the divisor. We go for a solution, where each pizza is cut into as many as pieces as the number of persons (divisor). If the number of pieces is the same as the number of persons, distribution is possible. The number of pizzas is taken as the result of division, which need not be a whole number. We get a quotient for the division problem. However, the next point is to compare the results of such two divisions. To solve this problem, the number of pieces cut in each pizza is the multiple of persons for both problems (the product of both the divisors). Now, both distributions have pieces of the same size and such pieces can be added or compared. We have taken a pair of the number of pizzas and the number of persons as a division expression. The equivalence of division and the arithmetic of such expressions are done on the basis of the observations of pizza distribution example. We have defined the relation on ordered pairs of integer and the natural number as the rational number with equivalence. Operations are defined in order pairs [3]. It is established that the relation is an equivalence relation, and the operations preserve the equivalence relation. The partition defined by the equivalence relation is a set of rational numbers. With the operations of addition and multiplication on the set of rational numbers it is a totally ordered field. This field is good enough for most of the routine requirements. However, the same is not enough. Equations like $x^2 = 2$ do not have a solution. We have explained the decimal representation of rational numbers, though primarily they are defined as the ratio of integers. It is not very convenient to express very large and small numbers using either of the two methods. Then, floating-point representation is introduced. Even floating-point representation can take an arbitrary base. Base values of 2 and 16 are popular in computers. The exact implementation may depend on the implementation standard. We move to the extension of rational numbers to real numbers in the next chapter.

Rational numbers do not have a square root of numbers, which are not complete squares. We need to extend the set of rational numbers to overcome this shortcoming. We address a broader shortcoming of the set of rational numbers not having roots of some polynomial equations. The set of rational numbers does not have the least upper bound property. There are some of the bounded above subsets, which do not have the least upper bound in the set of rational numbers. We extend the set of rational numbers to the set of Dedekind cuts. For better understanding, we have introduced rational cuts. The set of rational cuts represents rational numbers in the terminology of cuts. Though this does not extend the set of rational numbers, it paves a path to a better understanding of Dedekind cuts. Key properties of rational cuts are identified, and then Dedekind cuts are defined [4]. We prove that the set of Dedekind cuts with binary operations is a field, which is an extension of rational numbers and has the least upper bound property. Moreover, it is an ordered field. It is established that the set of real numbers is an uncountable infinite set. As the set of real numbers does not solve equations, where the square of the unknown is negative. To overcome this shortcoming, we define complex numbers in the next chapter.

We define the set of complex numbers as ordered pairs of real numbers, where each ordered pair is distinct. Addition and multiplication are defined on ordered pairs. We prove that the set of ordered pairs of real numbers with binary operations is a field. We prove that a subset of complex numbers is equivalent to the set of real-number-preserving operations. We establish that there is a complex number, which cannot be expressed as a linear combination of complex numbers equivalent to real numbers. The unit complex number of this type gives -1 when it is squared. Thus, we have found the solution of the square root of negative numbers. We have explained the Cartesian representation of complex numbers. Polar representation is introduced, and simplified formulae for multiplication, power, and roots of complex numbers are explained. Conjugate of the complex number is introduced. It is further established that the field of complex numbers is not ordered. We have explained that the cardinality of the set of complex numbers and that of real numbers is the same.

We hope the readers would enjoy a wonderful journey from natural numbers to complex numbers.

REFERENCES

1. I. N. Herstein, *Topics in algebra*, John Wiley & Sons, 2006.
2. K. H. Rosen, *Discrete mathematics & applications*, McGraw-Hill, 2012.
3. H. E. Campbell, *The structure of arithmetic*, New York: Appleton-Century-Crofts, 1970.
4. W. Rudin, *Real and complex analysis*, Tata McGraw-hill education, 2006.

REFERENCES

Author biographies

Prof. Nita received her PhD in statistics from Gujarat University in 1994. From February 1990 until now, she is the HOD of the Department of Mathematics at Gujarat University, India. She is a post-doctoral visiting research fellow at the University of New Brunswick, Canada. Prof. Nita's research interests include inventory modeling in the supply chain, robotic modeling, mathematical modeling of infectious diseases, image processing, dynamical systems and its applications, etc. Prof. Nita has published 13 monographs, 5 textbooks, and 475+ peer-reviewed research papers. Four edited books were prepared for IGI-Global and Springer, with a coeditor. Her papers are published in high-impact Elsevier, Inderscience, and Taylor and Francis journals. She is the author of 14 books.

Dr. Thakkar completed his masters in mathematics from Gujarat University, India, and was the recipient of the National Merit Scholarship during his post-graduation study. The study was followed by 35+ years of experience in information technology in the area of application solution design, development, and implementation, including ERP implementation and real-time process monitoring. After retirement from the industry, he completed his PhD in mathematics under the guidance of Prof. (Dr.) Nita H. Shah.

Natural Numbers 1

Our introduction to mathematics starts with positive whole numbers, also known as natural numbers. We will be going through informal and formal aspects of all types of numbers. We know that natural numbers are not enough for mathematics. For formal aspects of the development and extension of number types, we need background knowledge of some topics in mathematics. Hence, we start with the introduction of these concepts.

1.1 PREREQUISITES

1.1.1 Set Theory

We start with definitions related to sets. First of all, we define set. The definition given here is intuitive and not a formal one. The formal definition may involve some other terms, which may in turn require the definition of those terms. So we start with set as the initial term with an intuitive definition.

Definition 1.1: *Set* is an unordered collection of distinct objects.

For a collection being considered to be qualified as a set, one should be able to check unambiguously with certainty whether any object is in the collection or not. Generally, sets have the same type of members, but it is not necessary to be so. It may contain different kinds of objects. A common convention is to represent a set by an uppercase letter, and the member is represented by a lowercase letter. Lowercase letters are used as a member variable symbol, and the actual member can be anything. Using a lowercase letter to represent the member does not mean that members are lowercase letters. To indicate that an object a is in set P, we write $a \in P$, read as a belongs to P or a is in P. Similarly, if b is not in P then we write $b \notin P$, read as b does not belong to P or b is not in P.

One of the methods to describe a set is listing all its members. If the number of members is large, then ... is used to indicate terms similar to listed near it. The method of describing a set using a member list is called *roster* method. Another method of describing set is by *set builder*. In this method, a representative member is written on the left side of the vertical bar I, and related details are written on the right side of the vertical bar. In both the methods, set details are enclosed in curly brackets {}.

Some popular sets are: set of natural numbers denoted by N or $\mathbb{N}$, integers denoted by Z or $\mathbb{Z}$, rational numbers denoted by Q or $\mathbb{Q}$, real numbers denoted by R or $\mathbb{R}$, and complex numbers by C or $\mathbb{C}$.

Definition 1.2: Set A is a *subset* of set B if and only if all elements of A are elements of B.

Set A is a subset of itself as all members of A are in it. The symbol $\subseteq$ used to indicate that set is subset of another set. To indicate that the subset is not the set itself, symbol $\subset$ is used. If the subset being set itself is not significant, the symbol $\subset$ is used as a common symbol.

For a set to not become a subset of another set, it should have an element, which is not in the other set. For set A to not become a subset of B, there should be an element $a \in A \ni a \notin B$.

An alternate representation of 'A is subset of B' ($A \subseteq B$) is 'B is superset of A', denoted as $B \supseteq A$.

An important point to be noted here is two sets A and B are equal if and only if $A \subseteq B$ and $B \subseteq A$.

Definition 1.3: The set containing no elements is called an *empty set* or *null set*.

The null set is well-defined. Any candidate object can be checked for being a member of this set. Given any candidate, the result for membership of the set is that it is not a member. The symbol used for the null set is ϕ.

The null set does not have any element. So, for another set A, null set having an element which is not in set A is not possible as the null set has no element in it. Hence, the null set is a subset of all the sets.

An important point to note here is that the null set and the set itself are always subsets of any set.

Definition 1.4: A subset of a given set, other than the null set and the set itself, is a *proper subset* of the set.

We have defined a set containing nothing in it (an empty collection). Can we define a set that contains everything? We get tempted to believe that such a set can be defined. Unfortunately, we cannot define such a set. Definition of such a set leads to a paradox. Hence, we need to define a universal set for the context in which we are working. There is no universally universal set.

Definition 1.5: A set containing all objects, relevant to the context, is a *universal set*.

In the context, any set being considered is always a subset of a universal set.

Definition 1.6: For the given two sets, a set containing elements, which are in any of the two sets, is the *union* of the sets.

Union operation of sets is denoted by symbol $\cup$ between set symbols.

Definition 1.7: For an arbitrary collection of sets, the *union of the collection of sets* is a set containing elements, which are in any of the sets in the collection of sets.

Collection of sets is represented as $C = \{A_\alpha | A_\alpha$ is set to be included in collection$\}$, where C is arbitrarily selected symbol for collection set, arbitrary symbol A for set letter subscripted by arbitrary Greek subscript letter α represents member of collection.

Definition 1.8: For the given two sets, a set containing elements, which are in both the sets, is the *intersection* of the sets.

Definition 1.9: For the arbitrary collection of sets, the *intersection of the collection of sets* is a set containing elements, which are in all of the sets in the collection of sets.

Definition 1.10: Set of all subsets of a set is the *power set* of the given set.

The symbol used to indicate a power set is $\mathcal{P}$. The set of all subsets of set A is called the power set of set A and is written as $\mathcal{P}(A)$.

An important point to note here is that if the number of elements in a set is n, then the number of all subsets is 2^n.

Definition 1.11: A set of all ordered pairs, having the first element in ordered pair from the first set and the second element in the ordered pair from the second set is called *Cartesian product* or simply the product of the sets.

The symbol used to represent the Cartesian product of sets A and B is $A \times B$. Symbolically, a product set is defined as $A \times B = \{(a,b) | a \in A, b \in B\}$. The fact about the number of elements in a product set is that if set A has m elements and set B has n elements, then $A \times B$ has $m \times n$ elements.

1.1.2 Relation

Definition 1.12: A subset R of a product set $A \times B$ is a *relation* from A to B.

If the product set is $A \times A$ then it is called relation in A.

If an ordered pair (a,b) belongs to relation R, then we write it as aRb and read it as a is related to b. If an ordered pair $(a,b) \notin R$ then we write it as $a\bcancel{R}b$ and read it as a is not related to b. An important point to be noted here is that the relation has to be any subset of the product set, and there need not be a rule, though in most of the cases there may be some.

Example 1.1: $N = \{India, Sri Lanka, Nepal\}$ $C = \{(Kathmandu, Delhi, Colombo, Paris, Bombay, Rome\}$. Relation P from N to C is $\{(India, Delhi), (Sri Lanka, Colombo), (Nepal, Kathmandu)\}$. We can name the relation as country to capital. If we reverse the order in pairs, the relation becomes from C to N, capital to country. Another relation could be some of the countries to cities in each of them. We can list this relation as $\{(India, Delhi), (India, Bombay), (Sri Lanka, Colombo)\}$.

Now, we focus on special cases, where the relation is in a set (subset of $A \times A$ for some set A).

Definition 1.13: A relation R in set A is *reflexive* if $aRa \ \forall \ a \in A$.

Example 1.2: A is set of some persons. Relation S is a set of ordered pairs of persons having the same parents. The condition of having the same parents is satisfied for each person with one's self. This relation is reflexive.

Definition 1.14: A relation R in set A is *symmetric* if $\forall aRb, bRa$ holds good.

The relation described in Example 1.2 is symmetric as well.

Definition 1.15: A relation R in set A is *transitive* if $\forall aRb$ and bRc, aRc holds good.

The relation described in Example 1.2 is transitive as well.

Definition 1.16: A relation in set A is an *equivalence relation* if it is reflexive, symmetric, and transitive.

The relation described in Example 1.2 is an equivalence relation.

Example 1.3: For any set A, $A \times A$ is an equivalence relation.

Example 1.4: For any set A, relation $I = \{(a,a) \mid a \in A\}$ is an equivalence relation.

Example 1.5: Relation in the power set of a non-empty set A, $_BR_C$ if and only if $B \subseteq C$ is reflexive and transitive.

Definition 1.17: For an equivalence relation R in set A, set $\{b \mid aRb\}$ is the *equivalent class* for member a of set A.

The equivalent class generated by relation R for member a is denoted by $[a]_R$, which can be abbreviated to $[a]$ if the relation is unambiguous in the context.

It is obvious, and one can easily prove that for $a \in A, [a] \neq \phi$.

Theorem 1.1: For an equivalence relation R in set A for every member of set A, there exists a nonempty equivalent class.

Proof: For every $a \in A$, aRa as equivalence relation R is reflexive. So $a \in [a]$. Hence, $[a] \neq \phi$.

Theorem 1.2: For an equivalence relation R in set A, if $a, b \in A$ and $a \neq b$ then either $[a] = [b]$ or $[a] \cap [b] = \phi$.

We leave the proof to the reader.

Definition 1.18: Collection A_α of pairwise disjoint subsets of set A is such that union of those subsets is the set A then the collection of such subsets is *partition of the set A*.

Collection of equivalent classes for an equivalence relation in a set gives partition of the set.

For set $P = \{A_\alpha\}$ partition of set A, $R = \underset{A_\alpha \in P}{\cup} A_\alpha \times A_\alpha$ is an equivalence relation. We leave the proof of the statement to the reader.

Definition 1.19: A relation R in set A is *antisymmetric* if $\forall a, b \in A$, if $(a,b) \in R$ and $(b,a) \in R$ then $a = b$.

A simple interpretation of this definition is for distinct members a and b of set A; both (a,b) and (b,a) cannot be members of relation R for the relation to be antisymmetric.

Definition 1.20: A relation that is reflexive, antisymmetric, and transitive is a *partial order*, and the set in which the relation is defined is a *partially ordered set* with ordering defined by the relation.

Another term used for the partially ordered set is *poset*.

An important point to be noted here is that it may so happen that for partial order R, neither aRb nor bRa is true for some distinct members a and b of the set on which the relation is defined. This is the reason for it being called a partial order.

Example 1.6: For a given set A, if we define relation as being subset in the power set of A, then the relation is partially ordered relation.

Example 1.7: In the set of positive integers, if we define relation as being multiple of the number, then the relation is a partial order.

Definition 1.21: A partial order R defined on set A is *total order* if for each pair of distinct members a and b of set A either aRb or bRa.

We can easily check that partial orders defined in Examples 1.6 and 1.7 are not total orders.

Example 1.8: In a set of positive integers, relation $\leq$ (defined as aRb if $a \leq b$ is the total order).

Definition 1.22: For relation R, if $\exists b \neq c \ni aRb$ and aRc then the relation is said to have *one-to-many correspondence* or the relation is called *one-to-many relation*.

Example 1.9: Let us take set A of some countries and set B of some cities. If relation is defined as country to city in the country, it has one-to-many correspondence.

Definition 1.23: For relation R, if $\exists a \neq b \ni aRc$ and bRc, then the relation is said to have *many-to-one correspondence* or the relation is called *many-to-one relation*.

Example 1.10: Let us take set A of some cities and set B of countries. If the relation is defined as city to country to which it belongs, then the relation has many-to-one correspondence.

Definition 1.24: For relation R, aRb and $aRc \Rightarrow b = c$ for all members of relation, then the relation is said to have *one-to-one correspondence* or the relation is called *one-to-one relation*.

Example 1.11: Let us take set A of some countries and B of cities. If the relation is defined as country to its capital, then it has one-to-one correspondence.

Definition 1.25: For relation R, $\{a \mid (a,b) \in R\}$ is called the *domain* of the relation.

Definition 1.26: For relation R, $\{b \mid (a,b) \in R\}$ is called the *range* of the relation.

Definition 1.27: For relation R from A to B, B is the *co-domain* of the relation.

Definition 1.28: For relation R from A to B has a member (a,b) in it; then a is called *preimage* or *argument* (of b) and b is called *image* (of a).

1.1.3 Function

Definition 1.29: For given sets A and B, a subset R of $A \times B$ (a relation) is a *function* if $\forall a \in A$, $\exists (a,b) \in R$; moreover $(a,b) \in R$ and $(a,c) \in R \Rightarrow b = c$.

We can put the definition of function in simple words as it is a relation having the source set as the domain of the relation and does not have one-to-many correspondence.

Unlike relations, functions use a lowercase symbol followed by colon, domain set and co-domain set with an arrow from domain set to co-domain set, followed by the formula to find the image of the argument.

An important point to be noted here is that the function is not just a formula. The domain and co-domain are equally important.

Example 1.12: Let us take relation from Z (set of integers) to Z relating to square of argument integer. In set theory–based notations, the relation R is defined as a subset of $Z \times Z$, as $R = \{(a, a^2) \mid a \in Z\}$.

In notations popular for functions, the same is written as $f: Z \rightarrow Z f(x) = x^2$. The lowercase letter f is selected arbitrarily, and the image is expressed as a formula using pre-image.

Definition 1.30: For given sets A and B, a subset R of $A \times B$ (a relation) is a *partial function* if for $\forall (a, b) \in R$ and $(a, c) \in R \Rightarrow b = c$.

Put in simple words, it is a function, which need not be defined for all members of A.

Definition 1.31: If function $f: A \rightarrow B$, $f(a) = f(b) \Rightarrow a = b$, then the function is *one-one*.

The correspondence of the function is defined the same way as that of relation. Obviously, a function cannot have one-to-many correspondences.

Definition 1.32: For a function $f: A \rightarrow B$, if the range of function f is B, then the function is *onto*.

Definition 1.33: A function $f: A \rightarrow B$ which is not onto is *into*.

Definition 1.34: A one-one and onto function is called *bijection*.

Definition 1.35: If two functions $f: A \rightarrow B$ and $g: B \rightarrow C$ are defined, the *composite function* $g \circ f : A \rightarrow C$ is defined as $g \circ f(x) = g(f(x))$.

The composite function $g \circ f$ is read as g of f.

An important point to be noted is that it may so happen that $g \circ f$ is defined but $f \circ g$ is not defined. Even if both composite functions are defined, it is not necessary that they are the same.

Definition 1.36: For a set A, function $f: A \rightarrow A$ defined as $f(x) = x$ is the *identity function* on A. It is usually denoted as I_A.

A point to be noted here is for function $f: A \rightarrow B$, $I_B \circ f$ is the same function as f and $f \circ I_A$ is the same function as f.

Definition 1.37: For a function $f: A \rightarrow B$, if we can define function $g: B \rightarrow A$ such that $g \circ f = I_A$, then the function f is said to be *invertible* and the function g is called *inverse function* of f. Inverse function of function f is denoted as f^{-1}.

1.1.4 Cardinality

The first thing that comes to our mind when we come across a set is the number of elements in it. If we start listing elements of a set and if the list covers all elements, then we can find the total number of elements in the set.

Definition 1.38: If all the elements can be listed then the set is called a *finite* set.

Example 1.13: Set $\{1,3,5\}$ has three elements.

Example 1.14: We cannot count the number of elements in set $\{n|n$ is positive integer$\}$ because if we try to list members of the set, the list never ends. Hence, we cannot find the number of elements in this set.

Definition 1.39: For two sets A and B, if there exists a bijection (one-one and onto function) from A to B, then they have the same *cardinality*.

According to Schröder–Bernstein theorem, if for sets A and B, if there exists a one-one function from A to B and another one-one function from B to A then there exits a bijective function from A to B.

We write set symbol between two vertical bars (for example, we use $|A|$ for set A) to express cardinality of set.

Let us take the family of sets $N_k = \{n \mid n$ is positive integer $\leq k\}$. We can consider cardinality of each set $|N_k| = k$.

For sets in which elements get exhausted if we try to list, cardinality is the number of elements in it.

Definition 1.40: Relation $\leq$ is defined in the set of cardinalities of set of sets as $|A| \leq |B|$ if $\exists$ a one-one function from A to B.

The relation defined here is the total order in cardinality of sets.

Definition 1.41: A set is *infinite* if there exists a bijection from it to its proper subset.

An alternate definition of the finite set is that if a set is not an infinite set then it is a finite set. One can prove the equivalence of both the definitions.

Definition 1.42: A set which is either finite or has cardinality as same as that of the set of natural numbers is a *countable set*.

Some people consider only sets with cardinality as that of the set of positive integers as countable sets. To make things unambiguous, we fully qualify countable sets as countably finite or countably infinite set.

We can list elements of a countably infinite set as the sequence of members like $a_1, a_2, a_3, \ldots$

We can list elements of a finite set as terms of finite sequence like $b_1, b_2, \ldots, b_n$.

Elements of the union of a finite set $\{b_1, b_2, \ldots, b_n\}$ with the countably infinite set $\{a_1, a_2, \ldots\}$ can be expressed as sequence $b_1, b_2, \ldots, b_n, a_1, a_2, \ldots$ (repeating members can be dropped). The sequence is going to never end as even if all the members of the finite set are already there in the infinite set, infinitely many terms will be there in the union set.

The union of two infinite countable sets $A = \{a_1, a_2, \ldots\}$ and $B = \{b_1, b_2, \ldots\}$ is an infinite countable set. We can list members of both the sets as $a_1, a_2, a_3, \ldots$ and $b_1, b_2, b_3, \ldots$, respectively. We can write the members of union set as $a_1, b_1, a_2, b_2, a_3, b_3, \ldots$. This establishes that terms of the union of both the sets can be represented as terms of a sequence. So the union of two countably infinite sets is a countable (infinite) set.

Applying the above two results, the finite union of countable sets (with at least one infinite) is countable (infinite).

The set of all integers can be considered as the union of three sets, namely the set of positive integers, singleton set containing 0 and set of negative integers. Hence, the set of integers is countable.

Now, let us consider countably infinitely many countable (infinite) sets. All members of all sets can be written as doubly subscripted symbols with the first subscript as the set number and the second subscript as the element number in a set.

All the members are doubly subscripted. Now let us look at the sum of subscripts. The smallest value of such sums is 2, and then it takes all larger integers. If subscripts i and j are to be added up to m, then the subscript order pairs become $((1, m-1), (2, m-2), \ldots, (m-1, 1))$.

Consider the sequence with members listed with the sum of subscripts starting from 2 as explained above. This sequence covers all members of all the sets. The members which have already appeared earlier can be skipped. Thus, we have established that the countably infinite union of countably infinite sets is countable. This establishes that the set of rational numbers is countable. The union of collection of sets $Q_n = \left\{ \dfrac{m}{n} \mid m \text{ is integer} \right\}$, where n is positive integer gives the set of rational numbers. Thus, the set of all the rational numbers can be written as $\{q_1, q_2, q_3, \ldots\}$. Hence, rational numbers are countably infinite.

Some important results related to cardinality are listed below:

No set and its power set can have the same cardinality. This is proved by Cantor's theorem.

The symbol for the cardinality of an infinite countable set is $\aleph_0$ (read as aleph naught). For each set, its power set has higher cardinality. The power set of the countable set has next higher cardinality, and the same is written as $\aleph_1$. We can have such sequence of ordered cardinalities.

The set of real numbers has higher cardinality comparable to a countable set. In other words, the set of real numbers is a non-countable infinite set. It has a cardinality of $\aleph_1$. An unsolved problem is that "Is there a set having cardinality strictly between that of a countable set and the set of real numbers?".

1.1.5 Algebra

In normal algebra working, like factorization and equation solving, we use certain properties of operations such as addition and multiplication. For example, in solving $x - 7 = 3$ we add 7 to both the sides to make it $(x - 7) + 7 = 3 + 7$, which in turn gets transformed to $x + (-7 + 7) = 10$, which ultimately gives solution $x = 10$. Some people transfer -7 from the left side to the right side with a sign change and solve the equation. In this example, we have used the associative property of addition $(a + b) + c = a + (b + c)$. Two more concepts, existence of identity and inverse, are used, which go unnoticed.

In this section, we will study the formal aspect of different operations in algebra.

Definition 1.43: For a given set A, a function from $A \times A \to A$ is called a *binary operation* on set A.

For binary operation, operator symbol is used rather than usual function notation like $f(a,b)$. We use operator symbols like $+, -, \times, \div, \oplus, \otimes, \odot, \circ, *$ or $\wedge$. Any symbol, which does not create ambiguity can be used as binary operation symbol. Moreover, the operation is written as pair of the set members and the operation symbol between them. The only condition is that it must be defined for all ordered pairs of the set on which the operation is defined. The arguments of the function are called operands. Order of operands is important as a is first argument and b is second argument $f(a,b)$ whereas b is first argument and a is second argument in $f(b,a)$. As an example, we can take binary operation $\wedge$ on natural numbers defined as $a \wedge b = a^b$ has a as first operand and b as second operand. Changing order of operands, the operation gives different results.

Definition 1.44: For a given set A, a partial function from $A \times A \to A$ is *partial binary operation*.

Example 1.15: For set of positive integers, subtraction $-$ is a partial binary operation. For this operation, $5 - 2$ is defined but $2 - 5$ is not defined.

Example 1.16: For set of positive integers, division $\div$ is a partial binary operation. For this operation, $12 \div 3$ is defined but $3 \div 12$ is not defined.

Definition 1.45: For an operation $\oplus$ defined on set A if for $B \subseteq A$, if for $\forall a,b \in B$ if $a \oplus b \in B$ then operation $\oplus$ is *closed* in B.

Some people define *closure* property for the set on which it is defined. The operation being function, image of it (result of binary operation) is in co-domain, which is the set on which the operation is defined. Hence, closure of the operation is implied by definition of function.

Example 1.17: Addition operation defined on the set of integers is closed in a set of positive integers.

Example 1.18: Subtraction operation defined on the set of integers is not closed in a set of odd integers.

Definition 1.46: A binary operation $\oplus$ defined on set A is *associative* if $\forall a,b,c \in A \ (a \oplus b) \oplus c = a \oplus (b \oplus c)$.

We can do a grouping of operation arbitrarily (without changing order) if the operation is associative.

We can easily check that operation $\wedge$ defined as $a \wedge b = a^b$ is not associative as $(2 \wedge 2) \wedge 3 = 4 \wedge 3 = 64$. whereas $2 \wedge (2 \wedge 3) = 2 \wedge 8 = 256$. The operation is not associative as $(2 \wedge 2) \wedge 3 \neq 2 \wedge (2 \wedge 3)$.

Definition 1.47: A binary operation $\oplus$ defined on set A is *commutative* if $\forall a,b \in A, \ a \oplus b = b \oplus a$.

Definition 1.48: For a binary operation $\oplus$ on set A, if there exists a member e in A such that $\forall x \in A, \ e \oplus x = x$, then e is called the *left identity* for the binary operation; if $\forall x \in A, \ x \oplus e = x$, then e is the *right identity* for the binary operation; and if $\forall x \in A, \ e \oplus x = x$ and $x \oplus e = x$ then e is the *identity* for the binary operation $\oplus$.

Example 1.19: The set of positive integers N with binary operation addition does not have an identity.

Example 1.20: For binary operation multiplication defined on the set of positive natural numbers N, 1 is identity.

Example 1.21: The set of even positive integers with binary operation multiplication does not have an identity.

Definition 1.49: For a binary operation $\oplus$ defined on set A with additive identity e; for a member $a \in A$, if $\exists b \in A \ni a \oplus b = e$ and $b \oplus a = e$, then b is called the *inverse* of a, and it is written as $-a$ (for the operation being considered as some type of addition) or a^{-1} (for the operation being considered as some type of multiplication).

Similar to the left and right identify, left and right inverse can be defined.

Definition 1.50: Set A with binary operation $\oplus$ is called a *groupoid* or *magma*.

Definition 1.51: Set A with associative binary operation $\oplus$ is called *semigroup*.

Definition 1.52: Set A with partial binary operation $\oplus$ on it is called a *partial groupoid*.

Example 1.22: Set of natural numbers N with subtraction (–) as the binary operation is a partial groupoid.

Definition 1.53: Set A with associative binary operation having an identity element is called a *monoid*.

Definition 1.54: A set with associative binary operation having an identity element and having the inverse of each of its member is called a *group*.

An important point to be noted here is that for the group, the operation should be defined on the entire set, the operation should be associative, it must have an identity element, and each element must have an inverse element. It is not necessary for the operation to be commutative. Usually the symbol for set on which the operation is defined for the group is taken as G, and the operation symbol selected for it is one of $+, \times, \cdot, \circ$ or simply nothing. Binary operation expression is written as $a + b, a \times b, a \cdot b, a \circ b$ or simply ab, and the group is denoted by pair $(G, \oplus)$, where G is set on which binary operation $\oplus$ is defined.

Definition 1.55: If the binary operation of the group is commutative binary operation, then the group is called a *commutative group* or *Abelian group*.

Commutative groups are named Abelian groups in memory of great mathematician Niels Henrik Abel. The most famous Abel prize in the field of mathematics, popularly called Nobel Prize of Mathematics, is named in memory of the same mathematician.

Example 1.23: For a set A, the set of all invertible functions (one-one and onto) from set A to itself with composite as operation is a non-commutative group.

Definition 1.56: If two binary operations $\oplus$ and $\otimes$ are defined on set A satisfying $a \otimes (b \oplus c) = a \otimes b \oplus a \otimes c$ and $(b \oplus c) \otimes a = b \otimes a \oplus c \otimes a \ \ \forall a, b, c \in A$, then the operation $\otimes$ is said to be *distributive* over $\oplus$.

Individual sided distributivity can be defined on a similar line.

Example 1.24: For the power set of a given set A and binary operations $\cup$ (union) and $\cap$ (intersection), we can easily verify that both the operations are distributive over the other one.

Definition 1.57: A set A has two operations $\oplus$ and $\otimes$ defined on it. If the operation $\oplus$ forms a commutative group and $\otimes$ is associative, has an identity, and is distributive over $\oplus$, then it is called a *ring*.

The operations $\oplus$ and $\otimes$ are called addition and multiplication, respectively. Some mathematicians do not consider the existence of multiplicative identity necessary for the structure to become a ring.

Example 1.25: Set of integers with binary operations $+$ and $\times$ forms a ring. The set of integers is a commutative group with addition as the operation. Multiplication is associative and distributive over addition. It has got a multiplicative identity as well. Thus, it is a ring by both the definitions.

Definition 1.58: Set A has two operations $\oplus$ and $\otimes$ defined on it. If the operation $\oplus$ forms a commutative semigroup with identity and $\otimes$ is associative and distributive over $\oplus$, then it is called a *semiring*.

Definition 1.59: A ring (usually commutative ring) R with a total order $\leq$ is an *ordered ring* if it satisfies the following for every $a, b, c \in R$

1. If $a \leq b$ then $a + c \leq b + c$
2. If $0 \leq a$ and $0 \leq b$ then $0 \leq a \times b$, where 0 is the additive identity

Definition 1.60: A ring defined on set A with two binary operations $\oplus$ and $\otimes$ defined with an additional property of commutativity, the existence of identity (if not already covered in the ring), and inverse for all members except for additive identity for $\otimes$, then the ring is called a *field*.

Rearrangement of terms and taking out common can be done in the field. For example, $a \otimes b$ and $b \otimes a$ are equal in fields. Similarly, subtracting the same term from both the sides of the equation or dividing both the sides by non-zero expression can be done for equations in the field. In fact, in the routine calculation, it is done by changing the sign or dividing the other side when we move a variable (added or subtracted for the addition term transfer and divided when moved from the numerator) to the other side of the equation.

Example 1.26: Set Q of rational numbers with usual $+$ and $\times$ is a field.

Definition 1.61: A field F with the total order $\leq$ is an *ordered field* if it satisfies the following for every $a, b, c \in F$

1. If $a \leq b$ then $a + c \leq b + c$
2. If $0 \leq a$ and $0 \leq b$ then $0 \leq a \times b$, where 0 is the additive identity

For two algebraic structures, if there is a function from one to another preserving operation, then they are similar in a way.

Definition 1.62: Let there be groups $(G, \oplus)$ and $(H, \otimes)$. A function $f : G \to H$ satisfying $f(a \oplus b) = f(a) \otimes f(b)$ is called *Homeomorphism* from $(G, \oplus)$ to $(H, \otimes)$.

Definition 1.63: Let there be rings $R(\oplus, \otimes)$ and $S(\oplus, \circ)$. A function $f : R \to S$ satisfying $f(a \oplus b) = f(a) \oplus f(b)$ and $f(a \otimes b) = f(a) \circ f(b)$ is called *Homeomorphism* from ring $R(\oplus, \otimes)$ to $S(\oplus, \circ)$.

An important point to be noted here is that the operations are defined in the respective algebraic structure, and the same operator symbol does not mean that the operation definition is the same.

Definition 1.64: A homeomorphism, which is one-one and onto, is called *isomorphism*.

1.2 POSITIVE INTEGERS

1.2.1 Positive Integers in Real Life

Human beings start using mathematics with basic counting. In basic counting, some people include zero (0), whereas others start with 1. Those who include zero consider having nothing to be taken as zero (0) as the starting point, whereas others count if they have things, as according to them counting thing if we do not have does not have any meaning.

If we start counting, there is a starting point as described above. In either case, we can always add one more object and the count of objects is represented by the next number called a successor. If there is one object and if we add one more, the count of objects is represented by 2. The count before adding an object is called the predecessor of the count (after adding). In this example, 2 is the successor of 1 and 1 is the predecessor of 2. The number representing the starting point does not have a predecessor. We have defined a set of natural numbers using formal axioms subsequently in this chapter on these lines.

The process of addition is if one object is to be added, the result is the successor of the first number. If more than one object is to be added, then we add one (1) as the first step. The result after the first step is the successor of the first number, and the predecessor of the second number remains to be added. The process is done recursively. If a collection of m object and n object are combined into one collection, then the resultant collection has

$m+n$ objects. Adding the first collection to the second collection or the second collection to the first collection does not change the result. Hence, we can say that the addition of natural numbers is commutative (the order does not matter: $(a+b=b+a)$). Similarly, for three collections, whether we add the first two collections and then add third $(a+b)+c$ or we add the second and third collections and then add it to first $a+(b+c)$, the result remains the same. Hence, the addition is associative in positive integers.

Multiplication is taking collections, each having as many as the first operand objects; the number of collections taken are as many as the second operand. The result of multiplication, called the product, is the total number of objects. For multiplication, we take the collection of bricks. We put as many bricks as the first operand in a row and make as many rows as the second operand. The total number of bricks is the result of the multiplication. The shape formed by the arrangement is a rectangle formed with rows as the length side and rows forming width. If we look from the width side, rows become columns and columns become rows. This can be interpreted as the second operand multiplied by the first operand. This establishes that multiplication is commutative. If we need to multiply three operands, multiplication of the first two operands forms a rectangle. We can stack such rectangles (as many as the third operand). Total bricks in this structure gives the product of the three numbers as the first operand multiplied by the second and then multiplied by the third operand. We can look at the cuboid structure from different faces, and it can be interpreted as a different grouping of operands. Thus, we can say that multiplication is associative.

For the representation of positive integers, the most popular system is position value–based system. In decimal system, the rightmost position has a multiplier of 1, next on left has a multiplier of 10, and so on. We have explained this representation with two concepts. One of the concepts is using counter like the odometer of vehicle/electricity consumption meter, and another concept is making bundles of 10 of the objects and repeating the process to make larger bundles. We can take any positive integer >1 as the base. Due to the ease of representation and addition operation, computers use a *binary system*. In the binary system, the number of digits are only 2, making it simple to represent electronically. Result of addition has result digit and carry. Result digit and carry can be represented using Boolean algebra, which can be relatively easily implemented in electronic circuits.

If two different natural numbers are given and if we can count forward (by taking the successor every time repeatedly) to the second number, then the second number is greater than the first number. We can say that for positive integers a and b, $a+x=b$ has a solution for x if and only if b is greater than a (written as $b > a$) and the solution is written as $b - a$. The point to be noted here is that we can find $b - a$ if and only if $b > a$.

We have defined subtraction (with constraint). Now we attempt division. For the given natural number a, we can list its multiples $(a, 2a, 3a, ...)$. If an equation $ax = b$ is given with natural numbers a and b, then the equation has the solution for x if the value of b is a multiple of a. If b is the multiple of a then $b = ka$ for some natural number k. In such a case, k is the solution of the equation.

It is clear that there are cases, where a number is not divisible by another. We can understand this by a simple process of distributing objects. We cannot break the object. Let us have n objects, and we try to distribute them to m persons such that every person gets an equal number of the object. In the beginning, the number of objects to be distributed are n. If the number of objects is as many as the number of persons or more, then give one piece of object to each person. Now objects pending for distribution are $n - m$. We repeat the process for distributing pending objects. It is clear that after every distribution step, the number of objects pending for distribution reduces. At some point, the number of objects pending for distribution will be less than the number of persons. Now, no more distribution can be done. The remainder is the pending objects. We get a relation $n = q \times m + r$, where n is the number objects to be distributed to m persons such that all the persons get an equal number of objects, q is the number of objects each person has got, and r is the remaining objects, which is less than the number of persons. The value of q is zero or a positive integer.

With this background of positive integers, called natural numbers, we move to a formal definition of natural numbers, operations on them, and other properties of them.

1.2.2 Set Theoretic Definition of Natural Numbers

In this approach, the natural number 0 is defined as an empty set ϕ and for each natural number n its successor is defined as $n \cup \{n\}$.

This definition of natural numbers expands to $0 = \phi$, $1 = 0 \cup \{0\} = \phi \cup \{0\} = \{0\}$, $2 = 1 \cup \{1\} = \{0\} \cup \{1\} = \{0,1\}$, $3 = 2 \cup \{2\} = \{0,1\} \cup \{2\} = \{0,1,2\}$, $4 = 3 \cup \{3\} = \{1,2\} \cup \{3\} = \{1,2,3\}$....

So natural numbers can be listed as $0, 1 = \{0\}, 2 = \{0,1\}, 3 = \{0,1,2\}, 4 = \{0,1,2,3\}, \cdots$

If we try to remove recursion from the list, we can write natural numbers as ϕ, $\{\phi\}$, $\{\phi, \{\phi\}\}$, $\{\phi, \{\phi\}, \{\phi, \{\phi\}\}\}$ $\cdots$

This may sound too abstract to work with. Let us move to another axiomatic definition of natural numbers.

1.2.3 Peano Axioms

In this approach, natural numbers are defined using the following axioms.

1. 0 is a natural number.
2. For each natural number n, there is a unique successor $S(n)$, a natural number.
3. $S(n)$ is a one-one function. In other words, $m = n$ if and only if $S(m) = S(n)$.
4. For every natural number n, $S(n) = 0$ is false.
5. For set P, if $0 \in P$ and $n \in P \Rightarrow S(n) \in P \forall n \in P$, then set P contains all natural numbers.

To exclude 0 from natural numbers, 0 in the first axiom should be replaced by 1. Moreover, in axiom numbers 4 and 5, 0 should be replaced by 1 like in the first axiom.

From the first axiom, we can say that 0 is a natural number. From the second axiom, there is a successor for 0; we name it as 1, and 1 is another natural number. Applying the same axiom, there is a successor for 1 and the same is named as 2. Thus, commonly understood natural numbers are defined by these axioms. The fourth axiom says that 0 is the starting point.

Now, we move to the definition of operations on this set of natural numbers denoted as N or $\mathbb{N}$.

Definition 1.65: *Addition* operation + on set N of natural numbers (inclusive of 0) is defined as follows:

$a + 0 = a$

$a + S(b) = S(a + b)$ where S is successor function.

If we wish to exclude 0 from set N of natural numbers, the definition gets modified as follows:

Definition 1.66: *Addition* operation + on set N of natural numbers (excluding 0) is defined as follows:

$a + 1 = S(a)$ where $S(a)$ is successor of a.

$a + S(b) = S(a + b)$ where S is successor function

The formal definition given here can be easily related to our physical process of adding natural numbers. Start from the first operand, we go on counting forward as many times as the second operand to get the result.

We accept associativity and commutativity of addition without giving proof here.

The set of natural numbers N (both the variants, with and without 0) is a commutative semigroup with addition as binary operation.

If we take N with 0 in it, it is monoid (semigroup with identity) with addition operation.

Even after including 0 in natural numbers, it is not a group as there is no additive inverse for all elements except for 0.

Definition 1.67: *Multiplication* operation denoted by operator $\cdot$ on set N of natural numbers with 0 in it is defined as follows:

$$a \cdot 0 = 0$$

$a \cdot S(b) = a + a \cdot b$ where S is successor function.

Definition 1.68: *Multiplication* operation on set N of natural numbers without 0 in it is defined as follows:

$$a \cdot 1 = a$$

$a \cdot S(b) = a + a \cdot b$ where S is successor function.

Similar to addition, these definitions can be related to the physical process of multiplication.

Popular symbols used for multiplication operation are $\cdot, \times$ or nothing between them. For integer symbols (variables taking integer value), two symbols are written without anything between them. For example, to indicate multiplication of a and b, we write ab. Even $a \cdot b$ and $a \times b$ can be used. For literal values of natural numbers, $\times$ is unambiguous. For example, 7×5 clearly communicates multiplication of 7 by 5 without any ambiguity. If we use $\cdot$ for multiplication, there could be ambiguity when $7 \cdot 5$ is written; some people may consider it as 7, followed by decimal point further followed by 5, which is different from 7 multiplied by 5. Omitting symbol for multiplication and writing 7 and 5 without anything between them unambiguously means 75 (seventy-five) and not 7×5. However, wherever there is no ambiguity, we may use any of the three conventions for multiplication.

We accept associativity and commutativity of multiplication without giving proof here.

The set of natural numbers N with multiplication as the binary operation is monoid as 1 (the successor of 0 in case of natural numbers having 0 or the first natural number if 0 is not included in natural numbers) is the identity for multiplication.

If we look at both addition and multiplication, multiplication is both ways distributive over addition. Symbolically, $a \cdot (b+c) = a \cdot b + a \cdot c$ and $(a+b) \cdot c = a \cdot c + b \cdot c$.

1.2.4 Ordering in Natural Numbers

Definition 1.69: For natural numbers a, b (including 0), if we can find a natural number c (including 0) $\ni a + c = b$, then $a \leq b$ (read as a is less than or equal to b).

For the set of natural numbers without 0, the definition gets modified as follows:

Definition 1.70: For natural numbers a, b (excluding 0), if $a = b$ or if we can find a natural number c (excluding 0) $\ni a + c = b$ then $a \leq b$.

We can verify that this relation is reflexive, antisymmetric, and transitive. In addition, one of $a \leq b$ or $b \leq a$ is always true for all values of a and b. Hence it is the total order.

We can represent natural numbers on a line.

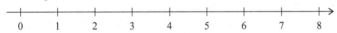

If we choose the option of natural numbers excluding 0, then the representation is as follows:

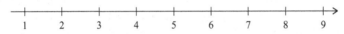

We can say that the numbers become greater as we move to the right.

Definition 1.71: Partial binary operation *subtraction* − is defined as $a - b = c$, if $\exists$ natural number $c \ni a = b + c$.

It is clear from the definition of $\leq$ that the subtraction operation is defined for the subtraction of a smaller number from a larger number.

Definition 1.72: For natural numbers a and b, if we can find $c \ni a = bc$ then $b \mid a$ (read as b *divides* a).

We can verify that this relation is reflexive, antisymmetric, and transitive. We can find examples where neither $a \mid b$ nor $b \mid a$ is true. A simple example is neither $2 \mid 3$ nor $3 \mid 2$. Hence, *divides* is a partial order but not a total order.

Definition 1.73: Partial binary operation division ÷ is defined as $a \div b = c$ if $\exists$ natural number $c \ni a = bc$.

It is clear from the definition of $\mid$ *(divides)* that a number can be divided only by its divisor.

1.2.5 First Principle of Mathematical Induction

A statement $P(n)$ is given for natural number n. If the following two statements are correct, then it is true for all natural numbers.

1. $P(1)$ is true
2. For natural number k, $P(k)$ is true $\Rightarrow P(k+1)$ is true

From Section 1.2.3 "Peano Axioms," it is clear that if the statement is true for each member of set A and $1 \in A$ and for $\forall k \in A \Rightarrow S(k) \in A$, where $S(k)$ is the successor of k then set A has all elements of N.

The first part can be changed to $P(0)$ to include 0 in the set, for which the statement is true.

Sometimes, a statement may not be true for some initial natural numbers. For example, the statement $4n < 3^n$ is true for all natural numbers starting from 2. In that case, we prove $P(2)$ is true as the first step in proof by mathematical induction.

Definition 1.74: A positive number greater than 1 is *prime* if it has no divisor except 1 and itself. It is called a *composite number* if it has a divisor other than 1 and itself.

The square of 1 is 1 itself. No other positive number has this property. For other primes, the square of the number is never itself. For other primer numbers, the square of it is not a prime. If we consider 1 a prime integer, then the special behavior (prime's square being prime) results in exception compared to other primes. Due to this behavior, many of the results regarding positive integers should have an overriding clause of ignoring 1 as prime. To overcome, we are excluding 1 from primes as well as composite numbers.

1.2.6 Second Principle of Mathematical Induction

A statement $P(n)$ is given for natural number n. If the following two statements are correct then it is true for all natural numbers.

1. $P(1)$ is true
2. For natural number k, $P(j)$ is true for $j \leq k \Rightarrow P(k+1)$ is true

Example 1.27: Every natural number ≥ 2 is either a prime number or the product of prime numbers.

The reader can prove this using the second principle of mathematical induction. The first part should be checked for 2, and the statement is to be proved for natural numbers ≥ 2. Statement is obviously true for 2. For the second part, the number could be prime or composite. In the case of composite, the assumption can be applied for both factors and the second part can be proven.

1.2.7 Well-Ordering Principle

Any non-empty subset of the set of natural numbers has the smallest element.

All the three principles (two of mathematical induction and well-ordering principle) are equivalent. The proof can be given, but the same is not included here.

1.2.8 Limitations of Natural Numbers

Two of the major shortcomings of natural numbers are listed below:

1. For arbitrary natural numbers a and b, finding natural number $c \ni a + c = b$ is not possible in all the cases. For example, for 2 and 5, $2 + c = 5$ is satisfied by $c = 3$. However, for 5 and 2, we cannot find natural number $c \ni 5 + c = 2$. As introduced in Section 1.2.4 (*Ordering* in Natural Numbers), for natural numbers a and b, $\exists$ natural number $c \ni a + c = b$ if $a \geq b$, but we cannot find $c \ni a + c = b$ otherwise. This holds good even if 0 is included in natural numbers. If 0 is excluded from natural numbers, $a + c = a$ does not have any solution for any value of a.
2. For arbitrary natural numbers a and b, finding natural number $c \ni a \times c = b$ does not necessarily have a solution. For example, for 3 and 12, $3 \times c = 12$ is satisfied by $c = 4$. However, for 12 and 3, we cannot find $c \ni 12 \times c = 3$.

We will overcome these limitations in subsequent chapters. We overcome the first limitation in the next chapter on integers, whereas the second limitation, which exists in the set of integers, will be resolved in a subsequent chapter of rational numbers.

1.2.9 Representation of Natural Numbers

We have worked on natural numbers in the abstract form. The numbers should be written in the physical form. There are different systems in use. Systems like Roman numerals (*I, II, III, IV, V, ..., X, ..., L, ..., C, ..., D, ..., M, ...*) are less popular but are in use. The most popular system is position value–based system with *ten* symbols. Here *ten* is the number of fingers (including thumbs) of both the hands for a normal human being. There is a reason for writing 10 in words as with any base system; the number of symbols is written as 10 in the respective representation. This system has a set of symbols used to write the number. The value of the symbol depends on its position. The system we normally use has symbols 0, 1, 2, 3, 4, 5, 6, 7, 8 and 9. In this system for number 265, 5 has value 5 itself whereas 6 has value 60 and 2 has value 200. In number 526, 6 has value 6 itself whereas 2 has value 20 and 5 has value 500. We can see here that the value associated to a digit depends on its position. In number 555, the value of the rightmost 5 is 5 itself whereas middle 5 has value 50 and the leftmost 5 has value 500. Even in this explanation, what is the interpretation of 50 and 500? Let us do a process to understand position-based value system.

Let us start with some quantum of objects, which can be counted. Meaning of countability here is that if we go on removing one object at a time, after some iterations, nothing would be left. The symbols (0, 1, 2, 3, 4, 5, 6, 7, 8 and 9) in the popular system are ten, so the system is called the decimal system. We form maximum number of groups, each having 10 objects (as many as the number of symbols we have selected). There are three possibilities.

1. No group is formed
2. Some groups are formed, and some objects are left out
3. Some groups are formed, and no object is left out

The number of objects left are translated to a symbol. Symbol 0 is used if no object is left out. If some objects are left out, then an appropriate digit is selected. The symbol selected is the least significant digit.

If no groups are there, then the process is ended. If there are groups, then the process is repeated for groups. From the groups, larger groups are formed, and after making larger groups, some group would be left out. The process would be similar to the one in the first step. The symbol assigned to left-out groups is put on the left to the digit put in the last step. At some stage, the process of making groups would be over as nothing would be left.

Let us take an example. We are given some quantum of objects. We group them in each group having 10 objects (//////////). The idea of explaining 10 by set of / characters is necessary as even 10 uses the concept of position value. In the process, we find that some groups are formed and six objects (//////) are left. Symbol 6 is selected. So 6 becomes the least significant digit. Now, we make larger groups from the groups. It so happens that some larger groups are formed and two groups (//) are left. Symbol 2 is selected to represent groups (on the left of least significant digits). Thus, the number becomes 26 so far. Now, we group larger groups into further larger groups. At this stage, no further groups are formed as only five larger groups are left. So symbol 5 is selected for this position. The number becomes 526. Now no more further grouping is required, the process is ended here. Another interpretation is that there are six single objects, two bundles each having ten objects (//////////) and five boxes each having ten bundles are there.

1.2.9.1 Hexadecimal System

The choice of ten symbols (as many as a normal human has total fingers – including thumbs) is arbitrary. We can select any number of symbols greater than 1. Let us take symbols 0 (Zero), 1 (One), 2 (Two), 3 (Three), 4 (Four), 5 (Five), 6 (Six), 7 (Seven), 8 (Eight), 9 (Nine), A (Abel), B (Baker), C (Charlie), D (Dog), E (Easy), and F (Fox). In all, there are 16 symbols. Now let us do the same process for the quantum of objects we had taken in the previous example. If we make groups of objects, each group having 16 objects, we would be left with some groups and Easy (////////////////) individual objects. This would get us digit E (Easy) in rightmost position. Now, we make larger groups, each having 16 small groups. In this step, two larger groups are formed and no smaller group is left. This gives digit 0 (Zero) in the immediate left position to the filled digits. The number so far becomes 0E. Now, we try to make the next larger groups from groups. There are only two groups; the next bigger groups are not formed. So these groups would be represented by digit 2, and the number becomes 20E. This can be interpreted as Easy single objects, no bundles having 16 objects and 2 boxes each having 16 bundles having 16 objects each. Those who are familiar with pieces – dozen – gross system should be able to relate this to it.

Mathematical interpretation of representation with position value is that the rightmost digit has a multiplier of 1 and as we move to left, for each shift multiplier gets multiplied by B, where B is the base of the system.

Another concept to understand the system is of counter, like odometer of vehicle or electrical consumption meter. It has circular disks (cylindrical ring) put next to each other. Each circular disk has the same set of symbols (the digits for the number system) on a curved surface. The first symbol in the digits is called Zero. There is a viewing window in the assembly, showing one digit from each disk. Initially, all the disks are aligned to have Zeroes in all the positions in the viewing window. Only the rightmost ring can be rotated to get the next symbol. Whenever, any disk gets zero after operation, it pushes the next disk (on left) to the next position. The symbols used are (0, 1, 2, 3, 4, 5, 6, 7, 8, and 9). Let us have a system with five disks. The initial display in viewing the window is 00000. After the first operation (increment), it becomes 00001. After the next operation, it becomes

00002. At one stage, it becomes 00009. Now, in the next operation, the symbol on the rightmost position becomes 0. So the disk on its immediate next position on the left gets a push and it gets 1 in the display window. So the viewing window gets 00010. Continuing the process further, we get 00011, followed by 00012, and so on. At some stage, it becomes 00019. Now, one more operation gets 0 in the right-most position. It gives a push to the next position, which is having 1. So the symbol at that position becomes 2 and the viewing window gets 00020. Continuing the process, at some stage it gets 00099. In the next step, the least significant position becomes 0, giving push to the next position, which is having 9 becomes 0. By the same mechanism, it gives a push to the next position having 0 in it, getting 00100. The viewing window shows natural numbers one after another (starting with 0 and the displaying successor after each number). The number of symbols on disk can be any positive integer >1. An important point here is that those many unique symbols should be used. This assembly gives a number in the corresponding base.

The number, in base B system, is written using l digits as $d_l d_{l-1} d_{l-2} \cdots d_3 d_2 d_1 d_0$ is $\sum_{i=0}^{l} d_i B^i$ in summation notation. To find the representation of any number n in the system with base B, one has to find $d_i, i = 0,1,\ldots,l$, such that each $0 \le d_i < B$ and $n = \sum_{i=0}^{l} d_i B^i$.

Important point to be noted here is that one can add an arbitrary number of zeroes on the left side as this would not contribute anything in summation.

We use some facts about summation notation for an interesting and useful result.

$$\sum_{i=0}^{m \cdot k - 1} a_i = a_0 + a_1 + a_2 + \cdots + a_{m \cdot k - 1}$$

$$= (a_0 + a_1 + \cdots + a_{k-1}) + (a_k + a_{k+1} + \cdots + a_{2k-1})$$

$$+ \cdots + (a_{(m-1)k} + a_{(m-1)k+1} + \cdots + a_{mk-1})$$

$$= \sum_{j=0}^{k-1} a_j + \sum_{j=k}^{2k-1} a_j + \sum_{j=2k}^{3k-1} a_j + \cdots + \sum_{j=(m-1)k}^{mk-1} a_j$$

$$= \sum_{j=0}^{k-1} a_j + \sum_{j=0}^{k-1} a_{k+j} + \sum_{j=0}^{k-1} a_{2k+j} + \cdots + \sum_{j=0}^{k-1} a_{(m-1)k+j}$$

$$= \sum_{i=0}^{m-1} \sum_{j=0}^{k-1} a_{ik+j}$$

For a given value of l and selected value of k, let us select the smallest m such that $m \cdot k - 1 >= l$.

Number in the system with base B can be written as $\sum_{i=1}^{m \cdot k - 1} d_i B^i$, which can be rewritten as $\sum_{i=0}^{m-1} \sum_{j=0}^{k-1} d_{i \cdot k + j} B^{i \cdot k + j}$. By taking out the terms from the inner sum, which are independent of index, it becomes $\sum_{i=0}^{m-1} B^{i \cdot k} \sum_{j=0}^{k-1} d_{i \cdot k + j} B^j$.

If we take $D_i = \sum_{j=0}^{k-1} d_{i \cdot k + j} B^j$, the summation becomes $\sum_{i=0}^{m-1} D_i (B^k)^i$, which represents the number in the system with base B^k. This gives us the simple process of converting a number from base B to B^k and vice versa. Putting the process in simple words, to convert the number from base B^k, convert each of the digit in k – digit number in B base. The number $20E$ in the 16-digit system (known as the *hexadecimal system*) can be converted to 2-digit system (known as the *binary system*) as 0010 0000 1110.

If we wish the same binary number in an eight-digit system (known as the *octal system*), the base being 2^3, we group the binary number into three-digit groups (start grouping from the unit position). The number becomes 001 000 001 110, which becomes 1016.

One algorithm to evaluate the number, each digit is multiplied by the position value. In the case of octal number 1016, it gets evaluated to $1 \times 8^3 + 0 \times 8^2 + 1 \times 8^1 + 6 \times 8^0 = 1 \times 512 + 0 \times 64 + 1 \times 8 + 6 \times 1 = 526$.

If we take hexadecimal number $20E$, it gets evaluated to $2 \times 16^2 + 0 \times 16^1 + 14 \times 16^0 = 2 \times 256 + 0 \times 16 + 14 \times 1 = 512 + 14 = 526$.

If we take binary representation 0010 0000 1110, only the positions 1, 2, 3, and 9 (starting from the rightmost position and numbering from 0) are non-zero, the number gets evaluated to $2^9 + 2^3 + 2^2 + 2^1 = 512 + 8 + 4 + 2 = 526$.

Converting from our familiar base of 10 to any base B (positive integer > 1) divide the number by target base B; the remainder is the least significant digit. Now, convert the quotient using the same method. To convert 526 to octal, we divide the number by 8. Division gives quotient 65 and remainder 6. So the least significant digit is 6. Now, we work on 65. For 65, it gives quotient 8 and remainder 1. So the result so far is 16. Now we convert 8 to octal. Dividing it by 8 gives quotient 1 and remainder 0. The result so far becomes 016. Now, we have to convert quotient 1 to octal. Dividing 1 by 8 gives quotient 0 and remainder 1. Now the number becomes 1016. Now quotient being 0, the process is ended.

Operationally, we go on dividing the number as we do in factorization; the only difference is that we always divide by the base and put the quotient in

TABLE 1.1 Conversion of Positive Integers from Decimal System to Other Bases

TO HEXADECIMAL SYSTEM (BASE 16)				TO OCTAL SYSTEM (BASE 8)		
	526				526	
16	32	14		8	65	6
16	2	0		8	8	1
16	0	2		8	1	0
				8	0	1

Collecting digits from the top and putting from the right, we get 20*E*.

Collecting digits from the top and putting from the right, we get 1016.

the main column and in the additional column, we put the remainder. We are demonstrating the process in Table 1.1.

We can convert a number from the hexadecimal system to decimal, starting with 0 as the result. We multiply the result by base (16) and add the current digit value. We start processing digits from the leftmost position and shift to the right one place at a time. The first digit to be processed is 2. The initial result of 0 is multiplied by 16, and digit 2 is added to give 2. Now to process the next digit 0, we multiply the result by 16 and digit 0 is added to give 32. Now to process digit E, we multiply the result by 16, and by adding E (value 14), we get $32 \times 16 + 14 = 512 + 14 = 526$.

Similarly, when converting from octal 1016, we get $0 \times 8 + 1 = 1$, followed by $1 \times 8 + 0 = 8$; subsequently $8 \times 8 + 1 = 65$, and ultimately $65 \times 8 + 6 = 520 + 6 = 526$.

1.2.10 Number System Used by Computers

Computers use the binary system (base 2). The choice of base 2 is due to the ease of translating digits (0 and 1) into electronic circuits. Two different voltage levels, on-off position of the switch, magnetic field orientation, and diode status for current flowing can be mapped to binary digits (BInary digiT – BIT). Controlling these components and knowing status can be relatively easily done using digital electronics circuits. Moreover, basic arithmetical operation of addition can be easily done due to simplicity in terms of the relation between operands and result.

Addition of two numbers with all possible values is given in Table 1.2. It is clear the result digit is "exclusive or" of two inputs and carry is "and" of the inputs, if we map 1 and 0 to *true* and *false*, respectively. Building Boolean

TABLE 1.2   Result of Addition of Two Binary Digits

ADDITION			ADDITION (DIGIT)			ADDITION (CARRY)		
+	0	1	$+_d$	0	1	$+_c$	0	1
0	0	1	0	0	1	0	0	0
1	1	10	1	1	0	1	0	1

circuits is relatively easy in digital electronics. As a result, computers are built with the binary system as a core computing system.

Any number can be represented in the binary system provided a sufficient number of digits (size) are taken. For example, any non-negative integer up to decimal 1000 can be expressed using 10 *(ten)* binary digits. As the number of digits required is high for binary representation, it is not convenient to use the binary system for human communication. For example, expressing the number of order of 1000, it requires 10 binary digits. The number of digits required to represent the number in binary is approximately three times the number of digits required in the decimal system. The *octal* (base 8) and *hexadecimal* (base 16) systems use almost as many digits as the decimal system, and it is very easy to convert from these systems to the binary system and vice versa, as seen earlier in the previous section. Hence, octal and hexadecimal systems are very popular among computer professionals. Each family of computers has its own machine language. Many popular computer systems have the memory unit of 8 bits, called a byte. Some of the instructions are 1 byte long, having 2 bit operation code and 2 operands, each of 3 bits, which refer to 8 quick access memory locations, numbered from 0 to 7, called registers. If we look at the contents in the octal system, the most significant octal digit becomes the operation code and the next two digits are operands. Thus, in this situation, the octal system is very convenient for computer professionals to use.

We look at the memory location each having 8 bits. This type of memory unit can store an unsigned (non-negative) integer from 0 to 255 or two hexadecimal digits (0 to *FF*) or three octal digits, where the most significant digit can be between 0 and 3 and other two digits can have values from 0 to 7. Any decimal system digit (0 to 9) can be stored in 4 bits. So we can use 8-bit memory unit (location) to store two decimal digits. The system of storing the decimal digit in 4 bits is known as *Binary Coded Decimal (BCD system)*. Historically, some computers were using BCD system for processing. Modern computers support BCD, but computation is done using the binary system. We have some examples of BCD arithmetic in the context of addition being done in the binary system:

Example 1.28: Evaluate $12 + 13$ using BCD arithmetic.

$12 + 13 = (0001\ 0010)_2 + (0001\ 0011)_2$

=	0001 0010	Binary
+	0001 0011	Binary
=	0010 0101	Binary
=	25	BCD

Even a simple binary arithmetic addition gives the correct result.

Example 1.29: Evaluate $12 + 19$ using BCD arithmetic.

$12 + 19 = (0001\ 0010)_2 + (0001\ 1001)_2$

=	0001 0010	Binary
+	0001 1001	Binary
=	0010 1011	Binary
=	*2B*	BCD

The computer does addition using binary system. In this case, we get the unit digit, which is beyond the valid digit range. So, if we try to correct the result, we should transfer 10 from the unit digit to tens digit to get the correct result.

This example indicates that if the unit digit is greater than 9, 10 should be subtracted from it and tens digit should be incremented by 1.

With this correction after addition, we get the correct result 31 for our BCD arithmetic.

Example 1.30: Evaluate $18 + 19$ using BCD arithmetic.

$18 + 19 = (0001\ 1000)_2 + (0001\ 1001)_2$

=	0001 1000	Binary
+	0001 1001	Binary
=	0011 0001	Binary
=	31	BCD

If we look at the result, the result looks valid (digits are valid) but the result is not correct. There is a carry from the fourth bit from the right to the fifth bit (between BCD digits). This results in getting 16 removed from the unit digit part and 1 getting added to tens (which means 16 in binary). For *BCD arithmetic*, we need to get 10 reduced (and not 16) and 1 should be added in tens. The corrective action required here is to simply add 6 in the unit digit. After this correction, the result becomes 37, which is correct.

After each operation in the computer, the CPU has a status, which can be checked by a subsequent operation. When we do addition, there is one status available called half carry status. This indicates whether there was any carry between lower 4 bits to higher 4 bits. Using this if we intend to use BCD addition, we can use "add instruction," which does binary addition. This should be followed by decimal adjust addition instruction. This instruction checks if there is any half carry in the previous operation (binary addition) and increases the unit digit (least significant 4 bits) by 6. If there is no half carry, check the unit digit (least significant 4 bits). If the digit is above 9 then reduce it by 10 and add 1 in the upper 4-bit number. In other cases, this instruction does nothing. So modern computers support BCD arithmetic. To keep the explanation simple, we have skipped a similar phenomenon happening in a significant digit. CPU instruction makes sure that the BCD result is correct in all the cases, whether adjustment in result is required or not.

In computer arithmetic, predefined space (8 bits, 16 bits, 32 bits, or 64 bits are popular) is allocated for handling numbers. The result of the operation should be compatible with the size. If we add any number to the highest-possible number then the result cannot be accommodated in the memory used. A typical case is that adding 1 to the highest-possible number gives result 0 with an overflow error.

An important point to be noted here is that the computer interacts with humans using some computer program, and in most of the cases, these programs uses the decimal system for user interactions.

1.3 SUMMARY

1. Natural numbers are defined using axioms in two different ways.
2. Some mathematicians include 0 (Zero) in natural numbers, whereas others do not include it.
3. Fundamental definitions of binary operations, addition and multiplication, are done.
4. A partial order $\leq$ is defined for natural numbers, which is the total order.
5. First principle of mathematical induction: If a statement $P(n)$ for natural numbers is true for $n = 1$ and $P(k)$ is true $\Rightarrow P(k+1)$ is true, then the statement is true for all natural numbers.
6. Second principle of mathematical induction: If a statement $P(n)$ for natural numbers is true for $n = 1$ and $P(j)$ is true for $j \leq k \Rightarrow P(k+1)$ is true, then the statement is true for all natural numbers.
7. Well-ordering principle: Any non-empty subset of a set of natural numbers N has the smallest element.

8. Addition is commutative and associative. If 0 is included in a set of natural numbers then it has an additive identity. The additive inverse does not exist for all natural numbers (other than 0).

9. For natural numbers a, b, and unknown x, equation $a+x=b$ does not always have a solution for x in the set of natural numbers. In other words, subtraction is defined as a partial binary operation. It is defined only if the number to be subtracted is smaller than the other number.

10. Multiplication is commutative and associative. 1 is the multiplicative identity. However, the multiplicative inverse does not exist for all natural numbers other than 1.

11. For natural numbers a, b, and unknown x, equation $a \times x = b$ does not always have a solution for x in the set of natural numbers. In other words, division is defined as a partial binary operation. It is defined only if the divisor operand is the divisor of the dividend operand.

Integers

2

We have gone through formal mathematics of natural numbers. We exclude 0 from natural numbers for further working. We have already seen that both addition and multiplication are commutative and associative for natural numbers. Hence, terms can be rearranged and regrouped arbitrarily. Moreover, multiplication is distributive over addition.

2.1 INFORMAL INTRODUCTION OF INTEGERS

Natural numbers have got limited subtraction capability. We shall use it to extend natural numbers to integers. Let us consider a trading activity. A person buys something at some price called purchase price and sells the same at some price called sales price. To keep computation limited to natural numbers, we shall consider the amounts involved in trade as whole numbers in the currency of the place. The activity can either result in profit, loss or no profit and no loss. If the sales price is more than the purchase price then the trade results in profit and the profit amount is the purchase price subtracted from the sales price. If the purchase price is more than the sales price then the trade results in loss and the loss amount is the sales price subtracted from the purchase price. If both the prices are the same then it is neither profit nor loss. We represent trade as an ordered pair having sales price in first the position and purchase price in the second.

There are some interesting properties of trades. If we add a fixed amount in both sales and purchase price, profit or loss remains the same. If we add a fixed amount to the sales price of a trade resulting in profit, then the profit increases by the amount added to the sales price. Similarly, if we add a fixed amount to the purchase price of trade resulting in loss, the loss increases by the amount added in the purchase price.

We can consider trades giving the same profit as the equivalent. Similar to profitable trades, trades resulting in the same amount of loss are also equivalent.

Let sales prices and purchase prices of one trade be s_1 and p_1, respectively, and the same be s_2 and p_2 for a second trade. We would like to consider the trades as equivalent if they result in the same profit or the same loss or no profit or no loss.

If both the trades have the same amount of profits, then $s_1 - p_1 = s_2 - p_2$. As indicated earlier, if we increase the sales price by a fixed amount, the profit of trade increases by the amount added. If we increase the sales price of both the trades by $(p_1 + p_2)$ then the profit for both the trades will increase by $(p_1 + p_2)$ and the profit will remain the same. For these trades, the sales price of both the trades is $s_1 + p_1 + p_2$ and $s_2 + p_1 + p_2$, respectively, whereas the purchase price remains the same as the original. Profits for both the trades are $(s_1 + p_1 + p_2) - p_1$ and $(s_2 + p_1 + p_2) - p_2$, respectively. Simplifying both the profit expressions, they become $s_1 + p_2$ and $s_2 + p_1$, respectively. As profit for both the trades is the same, the equation becomes $s_1 + p_2 = s_2 + p_1$. For two trades resulting in equivalent profits, the sum of sales price of the first trade and the purchase price of the second trade should be equal to the sum of purchase price of the first trade and the sales price of the second trade.

Similarly, for trades resulting in loss, the condition remains the same for them to have the same amount of loss. The same holds good even for no-profit no-loss trades.

If we put this condition in ordered pair form, trades (s_1, p_1) and (s_2, p_2) are equivalent if $s_1 + p_2 = p_1 + s_2$.

If we combine the two trades into one (adding respective amounts) then the outcome is the addition of two trades. The total purchase price is the sum of purchase prices of both trades and the sum of sales price is the total sales prices of both trades. We can decide the overall profit or loss from the total sales and purchase prices. If both the trades have profits then the profit of the combined trade (addition of both the trades) is the total of individual profits. If both the trades result in a loss, then the overall loss is the addition of individual losses. If one trade results in profit and another in loss, and if the profit value is more than the loss value, then the combined trade has profit as the difference of profit and loss of the trades. Similarly, if the loss of a trade is more than the profit of other trade, then the combined trade has loss as much as difference between loss and profit of the trades being combined. If the profit of one trade is the same as the loss of the other trade, then the combined trade has no profit or no loss. We can combine trades in terms of ordered pairs as $(s_1, p_1) + (s_2, p_2) = (s_1 + s_2, p_1 + p_2)$.

If a trade has the same sales and purchase prices then combining this trade with any other trade does not change the outcome in terms of profit or loss. Putting this in the ordered pair form, we can say that any trade of type (a, a) is a *Zero* trade.

Can we combine a trade with another so that the profit does not change for the combined trade? The answer for this is that the second trade should have the same value for sales price and purchase price. Thus, trades having the same sales price and purchase price are *Zero* trades.

For a trade with purchase price p_1 and sales price s_1, if we do another trade with purchase price s_1 and sales price p_1, the combined trade has a total purchase price $p_1 + s_1$ and sales price $s_1 + p_1$. The addition of natural numbers is commutative, for combined trade sales price and purchase price are same. Hence, we can say that second trade is negative of first trade. We have negative trade for such trades by reversing sales and purchase price. Putting this in ordered pair form, we say that $(s,p) + (p,s) = 0$.

If we consider trades giving profit as natural numbers, we can say that as far as addition is considered, the addition of trades reflects in the addition of respective natural numbers. We have negative trades for such trades by reversing sales and purchase price. If one trade is having a profit (natural number) and another trade is having a loss (negative of natural number) then the combined trade (addition of natural number and negative of natural number) gives profit if the profit value is larger (natural number larger than the absolute value of the negative of the natural number), loss if the loss value is larger (absolute value of negative of natural number larger than the other natural number) and zero if both are the same (natural number and the absolute of negative of natural number are the same).

For multiplication, the multiplication of trades does not seem like a meaningful process. A meaningful multiplication would be by repeating identical trades. We introduce reversal (negative) of trade. If a trade is done with purchase price p and sales price s then we define reversal of trade as a trade with purchase price s and sales price p. As seen earlier, reversal of a trade is negative of the trade. Combining one trade with its reversal trade results in no-profit no-loss, which is equivalent to no trade. We can combine trades and their reversal trades. For a combination of several trades, we can have the net of all the trades in one direction. If the count of reversal trades is more than that of original trades, then we have effectively done as many reversal trades as the difference of count of trades. Similarly, if the count of original trades is more than reversal trades then effectively there as many original trades as difference of counts of trades. If all the trades are in one direction then combined trade in the same direction. If the count in both the directions is the same, then effectively there is no trade. If the original trades have a profit then the overall outcomes are as listed below:

1. If net trades are original trades (positive count) then the total profit is the product of per-trade profit and the overall count of trades (product of natural numbers).
2. If net trades are reversal trades (original trade has a profit) then the reversal trade has the same amount of loss per trade. So, overall,

there is loss, which is the product of per trade loss (per trade loss amount is the same as per trade profit of original trade) and the number of net reversal trades (again product of natural numbers). The outcome being loss, the amount is considered to be negative. If we look at the activity from the count perspective, the count of trades is negative (reversals being more) and each trade having profit (our reference is original trade), then profit is

$$(\text{negative count}) \times (\text{positive profit}) =$$

$$-(\text{positive net count} \times \text{positive profit value}).$$

3. If both the counts are the same, then the net count is 0. Hence, the total profit at the overall level is $0 \times$ per trade profit. On the other hand, original trades and reversal trades being the same, the profit of original trades and loss of reversal trades is the same. So, the overall profit is 0. Hence, (zero trades) $\times$ (per trade-positive profit) = 0.

If the original trade results in loss then the reversal trades will have the same amount of profit for each trade. The outcome of such trades will be as follows:

1. If net trades are original trades (positive count) then the total loss is the product of per product loss and net count of trades (product of natural numbers). However, loss is considered as a negative number. Hence, profit is

$$(\text{negative per trade profit}) \times (\text{positive count}) =$$

$$-(\text{per trade profit}) \times (\text{positive count}).$$

2. If net trades are reversal trades (negative count), where each trade (reversal of loss making trade) is making a profit then the overall outcome of activities will be (net count) $\times$ (profit per trade). We look at it from a signed numbers perspective, then the overall outcome is (negative count) $\times$ (negative per trade profit). Equating both, we get (negative count) $\times$ (negative profit) = (net positive count) $\times$ (per trade profit).

3. If the net count is 0, then the loss of original trades and profit of reversal trades become the same resulting in no-profit no-loss. Hence, (0 trades) $\times$ (negative profit per trade) = (0 profit).

In case of the original trade being no-profit no-loss, each of the detailed summary item and the net count of activities become 0. Hence, (0 trades) $\times$ (0 profit per trade) = (0 profit).

We can summarize the multiplication operation as under:

1. Multiplication of positive operands is as it is done in natural numbers.
2. Multiplication of one positive and one negative operand gives the negative of product of both the operands ignoring sign.
3. Multiplication of negative operands is the product of both the operands ignoring sign.
4. Multiplication of operands with one of them being 0 gives the product 0.

If m identical positive trades and n corresponding reversal trades are done then the total sales price of positive trades is sm and that of negative trades is pn, making the gross sales price $sm + pn$. Similarly the gross purchase price is $pm + sn$. The overall profit or loss can be determined by gross sales and the purchase price. In ordered pair form, the multiplication of the count of trades and trade prices can be represented as $(m,n) \times (s,p) = (ms + np, mp + ns)$.

If we summarize the trading model, we can say that for two given natural numbers, if the first natural number is larger than the second, it is equivalent to the natural number, which is the result of subtraction of a smaller number from a larger number; zero if both the natural numbers are the same and negative of natural number if the subtracted number is larger, which is the result of subtraction of the smaller number from the larger number. We have defined operations of addition and multiplication of integers expressed as pair of natural numbers.

We proceed to the formal study of this.

An interesting thing about the absence of additive inverse in natural numbers is that, though additive inverse does not exist, cancellation of the same additive term from both the sides is valid in natural numbers.

Theorem 2.1: For natural numbers a, b, and c, $a + c = b + c \Rightarrow a = b$.

Readers can prove this using the first principle of mathematical induction for all natural numbers denoted by c.

2.2 INTEGERS AS RELATION IN ORDERED PAIRS OF NATURAL NUMBERS

Let us define relation R in $N \times N$ as

$$(a,b)R(c,d) \Leftrightarrow a + d = b + c.$$

The reader should be able to relate this to the condition for equivalence of trades.

We can easily verify that the relation defined here is the equivalence relation.

Some interesting facts about this equivalence relation are listed below:

1. $(a,b)R(c,d) \Leftrightarrow (a+p,b)R(c+p,d)$.
2. $(a,b)R(c,d) \Leftrightarrow (a,b+p)R(c,d+p)$.
3. $(a,b)R(c,d) \Leftrightarrow (a+p,b+q)R(c+p,d+q)$.
4. $(a,b)R(a+p,b+p)$.

Readers should be able to relate this to the trading concept and all the facts can be proved very easily using the definition of relation and properties of natural numbers.

2.3 ORDERING IN ORDERED PAIRS

We define ordering relation $\leq$ as $(a,b) \leq (c,d) \Leftrightarrow (a+d) \leq (c+b)$.

Readers can easily verify that this relation is reflexive, antisymmetric, and transitive; hence it is partial ordering.

In addition, one of $(a+d) \leq (b+c)$ and $(b+c) \leq (a+d)$ is always true. Hence, one of $(a,b) \leq (c,d)$ and $(c,d) \leq (a,b)$ is always true. This combined with $\leq$ being reflexive, antisymmetric and transitive is total ordering.

2.4 OPERATIONS IN ORDERED PAIRS OF NATURAL NUMBERS

Definition 2.1: Binary operations *addition* $\oplus$ and *multiplication* $\otimes$ are defined as follows:

$$(a,b) \oplus (c,d) = (a+c, b+d)$$

$$(a,b) \otimes (c,d) = (ac+bd, ad+bc)$$

Readers should be able to relate this to combining of trades and multiplication of trades (a trade occurring c times and reverse of it occurring d times).

We can prove that the equivalence relation holds well on the operations as well. Thus, we have established that operations defined here maintain the relation.

2.5 PROPERTIES OF BINARY OPERATIONS

In this section, we are going to check the properties of binary operations. In this context, for equality of two members (a,b) and (c,d) to be equal, we are going to use the equivalence relation for equality $a+d=b+c \Rightarrow (a,b)=(c,d)$.

Using the definition of operations and equivalence and properties of natural numbers, we can easily prove that addition $\oplus$ is commutative and associative.

Theorem 2.2: $N \times N$ has additive identity (identity element for $\oplus$).

Proof: we claim that $\exists (a,b) \ni (c,d) \oplus (a,b) = (c,d) \forall (c,d) \in N \times N$. If such an ordered pair (a,b) exists then it is additive identity.

$$(c,d) \oplus (a,b) = (c,d)$$
$$\Leftrightarrow (c+a, d+b) = (c,d) \qquad \text{by definition of addition } \oplus$$
$$\Leftrightarrow c+a+d = d+b+c. \qquad \text{by equivalence relation definition.}$$

By rearranging terms and canceling the common additive term $c+d$ from both the sides using Theorem 2.1, condition in the last statement becomes $a=b$.

For any ordered pair (c,d) of natural numbers, $(c,d) \oplus (a,a) = (c+a, d+a)$ and as $c+(d+a) = d+(c+a)$, by commutativity and associativity of the addition in natural numbers. Hence, $(c,d) \oplus (a,a) = (c,d)$ for all natural numbers c, d, and a.

We can prove that $(a,a) \oplus (c,d) = (c,d)$ the same way or we can use commutativity of addition.

Hence, $(c,d) \oplus (a,a) = (a,a) \oplus (c,d) = (c,d)$ for all natural numbers a, c, and d. So, (a,a) is additive identity for all natural number values of a.

Theorem 2.3: For all ordered pairs (a,b) of natural numbers, additive inverse (inverse for $\oplus$) exists.

Proof: We know that any ordered pair of the form (c,c) is additive identity for any natural number c.

We claim that $\forall (a,b) \exists (p,q) \ni (a,b) \oplus (p,q) = (c,c)$.

$$(a,b) \oplus (p,q) = (c,c)$$
$$\Leftrightarrow (a+p, b+q) = (c,c) \qquad \text{by definition of addition } \oplus$$
$$\Leftrightarrow a+p+c = b+q+c \qquad \text{by definition of equivalence relation.}$$

Canceling the common additive term c from both the sides using Theorem 2.1, we get the condition $a+p = b+q$. If we take values $p=b$ and $q=a$, the condition is satisfied. As $(a,b) \oplus (b,a)$ evaluates to $(a+b, a+b)$, which is additive identity, (b,a) is the additive inverse of (a,b). Verification of the other side inverse is trivial.

Now, we check the properties of multiplication.

We can easily prove that multiplication $\otimes$ is commutative and associative using the definition of equivalence and properties of natural numbers.

Theorem 2.4: $N \times N$ has multiplicative identity (identity element for $\otimes$).

Proof: Let us assume that (a,b) is multiplicative identity.

For (a,b) to be a multiplicative identity $(c,d) \otimes (a,b) = (c,d)$ should hold good for all ordered pairs (c,d) of natural numbers.

$$(c,d) \otimes (a,b) = (c,d)$$
$$\Leftrightarrow (ca+db, cb+ad) = (c,d) \qquad \text{by definition of multiplication } \otimes$$
$$\Leftrightarrow ca+db+d = cb+ad+c \qquad \text{by definition of equivalence of relation}$$
$$\Leftrightarrow ca+d(b+1) = c(b+1)+ad \qquad \text{by distributivity of multiplication over}$$
$$\text{addition in natural numbers.}$$

The condition in the last statement can be satisfied for all possible values of c and d if $a = b+1$.

Any ordered pair of form $(n+1, n)$, where n is a natural number is a multiplicative identity. We can easily verify that $(a,b) \otimes (n+1, n) = (a,b)$. Verification of its being other side identity is trivial.

We can easily prove that multiplication is distributive over addition using definitions of addition and multiplication and equivalence relation and properties of natural numbers.

Theorem 2.5: All the ordered pairs of $N \times N$ do not have a multiplicative inverse.

Proof: It is sufficient to prove that $\exists (a,b) \in N \times N$ not having a multiplicative inverse. A point to be noted here is that such a pair should not be of additive identity.

We will prove that for ordered pairs of type $(a+k, a)$, where $k > 1$ does not have a multiplicative inverse. Additive identity is (a,a). Hence, $(a+k, a)$ is not an additive identity if $k > 1$. As we agree that additive identity need not have a multiplicative inverse, the element we have selected $(a+k,a)$ is not an additive identity.

For (c,d) to be multiplicative inverse of $(a+k,a)$, $(a+k,a) \otimes (c,d) = (n+1,n)$ should hold good for natural number n.

$(a+k,a) \otimes (c,d) = (n+1,n)$	condition for (c,d) to be multiplicative inverse of $(a+k,a)$
$\Leftrightarrow ((a+k)c+ad,(a+k)d+ac) = (n+1,n)$	by definition of multiplication $\otimes$
$\Leftrightarrow (ac+kc+ad,ad+kd+ac) = (n+1,n)$	by distributivity of multiplication over additon in N
$\Leftrightarrow ac+kc+ad+n = ad+kd+ac+n+1$	by definition of equivalence relation
$\Leftrightarrow kc+(ac+ad+n) = kd+1+(ac+ad+n)$	by rearrangement and regrouping of addition terms in N
$\Leftrightarrow kc = kd+1$	by cancellation of common additive term in N.

For any value of $k > 1$, the condition derived in the last statement cannot be satisfied. For example, for $k=2$, value on the left side is even whereas the value on right side is odd and they cannot be equal. Thus, the multiplicative inverse does not exist for all ordered pairs.

2.6 INTERPRETATION OF RELATION AND OPERATIONS

We have established that with the equivalence relation as equality and addition operation $\oplus$, $N \times N$ is a commutative group. Moreover, multiplication $\otimes$ is commutative, is associative, and has an identity. Additionally, multiplication is distributive over addition. Hence, $N \times N$ with equality and the defined operations of addition $\oplus$ and multiplication $\otimes$ is a ring.

This structure has additive identity and inverse, but it does not sound like the extension of a set of natural numbers. Considering the fact that the ordered pair (k,k) is the additive identity, all the ordered pairs $(a+k,b+k)$ are equivalent to (a,b). If we plot ordered pairs, with equivalent pairs using the same symbol, each diagonal line (line with 45° slope) has equivalent ordered pairs. If we extend these lines, they intersect with X-axis at natural numbers for some of them whereas others intersect on the left side of natural numbers on X-axis as illustrated in Figure 2.1.

Figure 2.1 shows that each equivalent line either intersects with the horizontal line with equation $y = 1$ or with the vertical line with equation $x = 1$. The extended line intersects the X-axis at predecessor of x value in case of intersection on the horizontal line, whereas it intersects the Y-axis at predecessor of y value on the negative side in the case of intersection on the vertical line. In case of intersection at the common point, the line intersects at the origin.

We understand this fact analytically now. For the ordered pair (a,b), exactly one of $a < b$ or $a = b$ or $a > b$ is true.

In case of $a < b \, \exists c \ni a + c = b$, where c is a natural number. Let us compare $(1, c + 1)$ and (a,b). For these to be true, $1 + b = c + 1 + a$ should be true. This gets

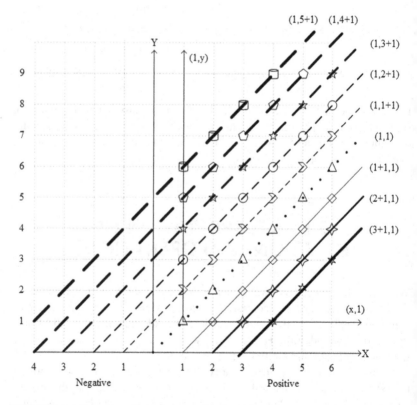

FIGURE 2.1   Plot of equivalent ordered pairs.

simplified to $a + c = b$. So, the ordered pair (a,b) becomes equivalent to $(1, 1 + c)$, where $c = b - a$.

Similarly, for $a > b$, (a, b) becomes equivalent to $(1 + c, 1)$, where $c = a - b$.
For $a = b$, (a,b) is equivalent to $(1,1)$.
All the ordered pairs can be represented as either $(a + 1, 1)$, $(1,1)$ or $(1, a + 1)$.

We can check the operation results in this standard format. The fact is that we can add or subtract the same term (if subtraction in natural numbers is defined) from both the co-ordinates.

$$(a + 1, 1) \oplus (b + 1, 1) = (a + b + 2, 2) = (a + b + 1, 1).$$

$$(a + 1, 1) \oplus (1, 1) = (a + 2, 2) = (a + 1, 1).$$

$$(a + 1, 1) \oplus (1, b + 1) = (a + 2, b + 2) = (a, b).$$

As discussed above, if $a > b$ then (a,b) is equivalent to $(a - b + 1, 1)$, whereas if $a < b$ then it is equivalent to $(1, b - a + 1)$. In the case of $a = b$ the result is equivalent to $(1,1)$. Considering commutativity the only case which remains to be checked is $(1, a + 1) \oplus (1, b + 1)$. $(1, a + 1) \oplus (1, b + 1) = (2, a + b + 2) = (1, a + b + 1)$

Now, we analyze the multiplication operation.

$(a + 1, 1) \otimes (b + 1, 1)$
$= ((a + 1)(b + 1) + 1, (a + 1)1 + (b + 1)1)$ by definition of multiplication $\otimes$
$= (ab + a + b + 1 + 1, a + 1 + b + 1)$ simplification using natural number arithmetic
$= (ab + 1, 1)$ elimination of $a + b + 1$ from both the components.

$(a, b) \otimes (1, 1)$
$= (a + b, a + b)$ by definition of multiplication $\otimes$
$= (1 + a + b, 1 + a + b)$ by adding 1 in both the components
$= (1, 1)$ by removing $a + b$ from both the components.

$(a + 1, 1) \otimes (1, b + 1)$
$= ((a + 1)1 + 1(b + 1), (a + 1)(b + 1) + 1)$ by definition of multiplication $\otimes$
$= (a + 1 + b + 1, ab + b + a + 1)$ by simplificaiton in natural numbers
$= (1, ab + 1)$ cancellation of common term $a+b+1$ from both the sides.

2.7 MAPPING OF ORDERED PAIRS AS EXTENSION OF NATURAL NUMBERS

For each ordered pair (a,b), exactly one of $a < b$, $a = b$ or $a > b$ is true. If $a < b$ then $\exists c \ni b = a + c$. Adding 1 to both the sides and rearranging the terms, we get $1 + b = 1 + c + a$. Using the definition of equivalence relation, $(1, 1 + c) = (a, b)$, where $c = b - a$. In this case, c is a natural number as $b > a$. Similarly, in the case of $a > b$, $(1 + c, 1) = (a, b)$, where $c = a - b$. In this case, c is a natural number as $a > b$. In case of $a = b$, by definition of equivalence relation, it can be equated to $(1, 1)$.

For the equivalence relation, there is a partition of equivalent classes. Partition for this relation $Z = \{[(1, 1 + a)] \mid a \in N\} \cup \{[(1, 1)]\} \cup \{[(a + 1, 1)] \mid a \in N\}$. This partition represents a set of integers. Symbol Z for the set of integer has its source in the German language phrase *ganze Zahlen*, literally meaning whole numbers.

Define a function from $A = \{(a + 1, 1) \mid a \in N\}$ to N as $f((a + 1), 1) = a$.

We can easily verify that the function is one-one and onto. As checked in the previous section

$$f((a + 1, 1) \oplus (b + 1, 1))$$

$$= f(a + b + 1, 1)$$

$$= a + b$$

$$= f(a + 1, 1) + f(b + 1, 1).$$

$$f((a + 1, 1) \otimes (b + 1, 1))$$

$$= f(ab + 1, 1)$$

$$= ab$$

$$= (f(a + 1, 1))(f(b + 1, 1)).$$

Using this, we can say that the subset $\{[(a + 1, 1)] \mid a \in N\}$ corresponds to natural numbers, the subset $\{[(1, 1)]\}$ corresponds to additive identity 0, and $\{[(1, 1 + a)] \mid a \in N\}$ corresponds to the additive inverses of natural numbers.

If we look at the operations, the addition of two positive integers gives the result as per the positive integer rules; negative numbers get added to the unsigned numbers with the negative sign. In natural numbers, the smaller integer can be subtracted from the large integer. For the addition of two numbers with different signs, the sign of result is that of the number with larger unsigned number and the value is smaller unsigned value subtracted from the larger unsigned value.

In multiplication, the result is 0 if any of the operands is 0, whereas for non-zero numbers, the result is the value of the product of unsigned numbers and the sign of the result is positive if both the operands have the same sign otherwise it is negative.

We have extended natural numbers to integers for having an additive identity and inverse preserving properties of natural numbers.

2.8 REPRESENTATION OF INTEGERS

Representation of integers is the same for positive integers. In the case of negative numbers, unsigned value with sign − is used. Separate sign indication is good for humans. The addition of numbers with the same sign is simple, but the addition of different signs requires special treatment. For computers, we are going to discuss an approach, where no separate treatment is required for negative numbers.

Let us revisit the counter concept. We go on incrementing the least significant digit. Each operation of increment gives the successor of the natural number in the display window. We can use decrement to find the predecessor value. We need to change the mechanism for decrementing digits to decrement the next position digit, when the current digit moves from 0 to the highest digit 9. As an illustration, let us start with 125 in the display window. Decrementing the least significant digit gives 124. Subsequently, we get 123, 122, 121, and 120. Now, when we decrement the least significant position digit, this gives digit 9 in that position. This transition from 0 to the highest digit, the number in next left position (with 2 as contents), should be decremented to 1. So, the number becomes 119. Repeating the process, at some stage, it becomes 100. Now the next decrement gets 9 in the least significant digit – a transition from 0 to 9 – resulting in decrement in the next position on the left. That position again has 0, which gets a new value 9 – again a transition from 0 to 9 – causing a decrement in next left position (having 1) gets 0 as its new value. This results in 099 in the display window. We can ignore the leading zero. If we repeat the process, at one stage we get 00000 in the display window. Now, one more decrement gets 9 in the least significant position. This triggers the decrement in the next left position changing the digit from 0 to 9. This happens for all positions, and the number becomes 99999. This gives an indication that 99999 can be considered as −1. Similarly, 99998, 99997, 99996, ⋯ can be treated as −2, −3, −4, ⋯, respectively. If we add a number to its negative number found by this method, the result becomes 100000, which gets transformed to 00000 as there are only five positions.

We can go by the paper-pen method to find the representation of negative numbers. For subtraction of a positive integer from another, we subtract the smaller number from the larger number to get the result value. If the number to be subtracted is larger, then the result is negative whereas it is positive otherwise. Let us experiment with the method of subtraction. Without comparing the numbers, we try to subtract the number. If the number to be subtracted is larger, then for the most significant position the digit will have trouble as we cannot subtract the most significant digit. We go for "borrow" from the non-existing digit. For finding the negative of 1, we try to subtract it from 0. We write both the numbers using the number of the pre-decided total positions we are going to use. If we have five positions, we try to subtract 00001 from 00000. For the least significant digit, we need "to borrow." After the process of borrowing, the least significant position gets an extra 10, whereas each of the other digits become 9. In traditional subtraction, this is not allowed as the most significant position 0 cannot borrow from the next significant digit as it is not there. We take an exception here and allow a "borrow" there. Now position wise subtraction gives a result of 99999. Effectively, we have subtracted 1 from 100000. Now, if we add 1 to its negative value 99999 (as decided by us), we get 100000. The result has only 5 digits. So, 1 in the most significant position is lost and the result is 0. This does not sound to be an acceptable concept for pure mathematicians. However, by this concept, computers can handle negative numbers very easily. If the largest unsigned number that can be handled is N then we compute representation of negative of number n as $N+1-n$. If we add the number n and its negative number representation $N+1-n$ the result becomes $N+1$, which gets truncated to 0. The biggest number that can be stored being N, we have to calculate $N-n$ and then add 1 to get an additive inverse (negative) of a number. In the system, the biggest number that can be stored is the largest digit in each position. Subtracting a number from it is subtracting each digit from the largest digit individually. The process of such subtraction is called complement. In the case of decimal system, the largest digit being 9, each digit is subtracted from it. This process is called taking 9s complement. For example, if we need to find the additive inverse of 5 (as a 5 digit decimal system number), we take the 9s complement for 00005, which is 99994 and then add 1 to it. We get the result 99995 at the end of the process. The result after adding 1 is called the *10s complement*. When we add 99995 to 5, the result becomes 100000 and the digit outside the storage capacity (1 in this case) is lost and the result becomes 0. So, we can treat 99995 as the additive inverse of 5. Now, if we try to find the additive inverse of 99995, its 9s complement is 00004. At the end, we add 1 to get the final additive inverse. Thus 5 becomes the additive inverse of 99995.

If we generalize the process, for a system with base B, having the largest digit L, to find the additive inverse, we take the Ls complement followed by the addition of 1 to the result. The final number is called Bs complement. In general, the additive inverse of a number is called the number's Bs complement.

For a decimal system, the additive inverse is called $10s$ complement and is found by adding 1 to the $9s$ complement. For the binary system, the additive inverse is called *2s complement* and is found by adding 1 to the $1s$ complement. The interesting thing about binary system is that the $1s$ complement is found by flipping the individual digit (BInary digit – called BIT).

The most significant digit represents the sign. In the most significant position, out of the digits in the number system, the smaller half values represent positive (non-negative as it includes 0) numbers, whereas the bigger half digits are for negative numbers. In computers, the entire system of interpretation is internal to the system. When it comes to interactions with humans, numbers are transformed to separate sign convention. For input, if an input number does not have a sign or has a positive sign, the same is converted to binary as the number is positive. In case of numbers with negative sign, unsigned numbers are converted to binary and then the $2s$ complement is stored as a value internally.

In the case of BCD arithmetic, as this is not primary data type handled by core instruction set, negative numbers are implemented as a separate sign number for the BCD arithmetic.

2.9 SUMMARY

We have extended the set of natural numbers to integers. The extension has the additive inverse of each of the members. Operations of addition and multiplication are extended. The set of integers with addition and multiplication becomes a group under addition and multiplication being associative and distributive over addition, is a ring. Multiplication operation being commutative, it is a commutative ring. Some people do not consider multiplication identity mandatory for the ring. If we go by this relaxation then this structure is a commutative ring with multiplicative identity.

Computers use two's complement to store a negative number.

Rational Numbers

3

We have gone through the properties of integers and operations on them in the previous chapter. An important thing about set of integers with the addition and multiplication operations is that it is a ring. Each integer does not have a multiplicative inverse, but multiplication cancellation is allowed in integers.

3.1 INFORMAL INTRODUCTION OF RATIONAL NUMBERS

We build an approach for the multiplicative inverse of integers. In this approach, we use a food object pizza. For this purpose, we will restrict our discussion to a positive number of pizzas and the number of persons among whom the pizzas are to be distributed. If the number of pizzas is multiple of the number of persons, we can distribute the pizzas such that every person gets some number of full pizzas (the number of pizzas divided by persons). If we do not want to go through division, we can go on giving one pizza to each person at a time, and after some iterations, all the pizzas would be distributed as the number of pizzas is multiple of the number of persons. The number of iterations of the distribution of pizzas (the number of pizzas each person has got) is the result of the division of the number of pizzas by the number of persons. Now, we attempt the process, which can be used in all cases.

Let us take the number of pizzas as z and the number of persons as n. To distribute, cut each one of them into n equal-size pieces (pies). Now, every single pizza can be distributed among n people as there are n persons and n pieces. As we distribute z pizzas by this process, each person gets z pieces (each piece being n^{th} part of a pizza). We represent this as an ordered pair having the number of pizzas as the first entry and the number of persons as the second entry. In this case, the ordered pair is (z, n).

We have an option of cutting each pizza into the multiple of n pieces. If we go by this approach, then if we cut a pizza into $n \times k$ pieces for some positive integer k, each person gets k pieces from each pizza. In both cases, the quantity of pizza got by each person is the same. This fact can be represented as (z,n) and (zk, nk) are equivalent.

There are two lots of pizzas and two groups of persons, where one lot of pizzas is to be distributed to the first group of people and the second lot of pizzas to the second group of people. If whole pizza distribution is possible in both cases, then if the first group of people has n_1 persons, the number of pizzas should be $k_1 \times n_1$ for some natural number k_1 and each person gets k_1 pizzas. Similarly, the second group has n_2 persons and $k_2 \times n_2$ pizzas and each person gets k_2 pizzas. If we add up pizzas received by each person in the first group and each person in the second group then it would be $k_1 + k_2$.

For the general case, if the first group has n_1 persons and z_1 pizzas are to be distributed and the second group has z_2 pizzas for n_2 persons, then each person of the first group gets z_1 pieces, where each piece is n_1^{th} part of a pizza, and each person of the second group gets z_2 pieces, where each piece is n_2^{th} part of a pizza. Pieces for the first and second groups are not comparable. To make both of them compatible, we cut each pizza into a common multiple of n_1 and n_2. A common multiple of both is $n_1 \times n_2$ pieces. The number of pieces for each pizza selected is not the minimum but can be easily selected. Now, if we distribute the pizzas in each of the group, each person of the first group gets n_2 pieces and each person of the second group gets n_1 pieces from each pizza. The quantity of pizza received by each person of the first group is $z_1 n_2$ pieces and each person of the second group is $z_2 n_1$ pieces, where each piece is $(n_1 \times n_2)^{th}$ part of a pizza. If we add up the quantity of each person from both groups, then it is $z_1 n_2 + z_2 n_1$ pieces, where each piece is $(n_1 \times n_2)^{th}$ part of a pizza. In ordered pair form, this can be represented as $(z_1, n_1) + (z_2, n_2) = (z_1 n_2 + z_2 n_1, n_1 n_2)$, as the computation of pieces shown above.

Putting the process in a symbolic form, if we distribute z pizzas among n persons, then each person gets $\dfrac{z}{n}$ (z divided into n) pizzas. This is an extension of the case, where the number of pizzas is the multiple of the number of persons. If two ratios $\dfrac{z_1}{n_1}$ and $\dfrac{z_2}{n_2}$ are taken, then they add up to $\dfrac{z_1 n_2 + z_2 n_1}{n_1 n_2}$.

We can introduce the multiplication of ratio by integer by taking pizzas as multiples of the basic distribution among n persons. If the original number of pizzas had been z and the number of persons n, then each person would get z pieces, where each piece is the nth part of a pizza. If we change the number of pizzas to zk, then each person gets zk pieces, where each piece is n^{th} part of a pizza. If we have original z number of pizzas for a multiple of n persons, say

ng persons, then each person gets z pieces of pizzas, where each piece is $(ng)^{th}$ part of a pizza. Now, if we change the number of pizzas to zk, then each person gets zk pieces of pizza, where each piece is $(ng)^{th}$ part of a pizza. By multiplying the number of pizzas by k, the result is multiplied by k, whereas multiplying persons by g, the result is divided by g. By combining both, we have multiplied by $\dfrac{k}{g}$. By the definition of multiplication, each person gets $\dfrac{z}{n} \times \dfrac{k}{g}$ pizzas, whereas by the explanation given above, each person gets $\dfrac{z \times k}{n \times g}$ pizza.

In ordered pair form, $(z, n) \times (k, g) = (zk, ng)$.

A trivial but important point to be noted is that if the number of pizzas is positive integer and persons is 0, then distribution such that nothing remains pending is not possible as all the pizzas remain pending. If both pizzas and persons are zero, then pizzas received by each person cannot be determined. If we take the number of pizzas as nk and persons as n then each person gets k pizzas. If we go on reducing pizzas by k and the number of persons by 1, even then each person gets k pizzas. At some stage, the number of pizzas becomes k and the number of persons becomes 1. At this stage, each person gets k pizzas. One more step of reduction results in 0 pizzas for 0 persons. Going by the same rule, each person in this situation gets k pizzas for any integer k. Hence, we cannot determine how many pizzas each person gets.

Now, we proceed to the formal definition of rational numbers, their equivalence, and operations on them on the basis of the understanding given above in Section 3.1, after stating important results for integers.

Lemma 3.1: For integers $a, b \in Z$, $ab = 0 \Rightarrow a = 0 \vee b = 0$.

Theorem 3.1: For non-zero integer a, $ab = ac \Rightarrow b = c$.

3.2 RATIONAL NUMBERS AS RELATION IN ORDERED PAIRS OF INTEGERS

Let us define relation R in $Z \times N$ as $(a, b)R(c, d)$, if and only if $ad = bc$.

An interesting thing about this relation is that of $(a, b)R(ap, bp)$, where p is a positive integer.

We can easily verify that R is an equivalence relation. We leave verification to the reader.

3.3 ORDERING IN ORDERED PAIRS

We define relation $\leq$ as an ordering relation as:

$$(a,b) \leq (c,d) \Leftrightarrow ad \leq bc.$$

We can verify that the relation is reflexive, antisymmetric, and transitive. Hence, the relation $\leq$ is a partial order.

In addition, for arbitrary ordered pairs (a, b) and (c, d), we can prove that one of $(a, b) \leq (c, d)$ and $(c, d) \leq (a, b)$ holds good. Hence, ordering $\leq$ is the total order.

3.4 OPERATIONS IN ORDERED PAIRS

We define two operations, addition $\oplus$ and multiplication $\otimes$, in ordered pairs of $Z \times N$.

Definition 3.1: Binary operations, *addition* $\oplus$ and *multiplication* $\otimes$, are defined as follows:

$$(a,b) \oplus (c,d) = (ad + bc, bd)$$

$$(a,b) \otimes (c,d) = (ac, bd)$$

We can prove that the operation preserves equivalence using the definition of operations, equivalence, and properties of integers. The reader should be able to relate this to the concepts introduced in pizza distribution.

3.5 PROPERTIES OF BINARY OPERATIONS

In this section, we are going to check the properties of binary operations $\oplus$ and $\otimes$ in ordered pairs of $Z \times N$.

The reader can easily prove that the addition is commutative and associative.

Theorem 3.2: Operation $\oplus$ has an identity (additive identity) in $Z \times N$.

We claim that $(0, q)$ is an additive identity for arbitrary natural number q.

We can easily verify that $(a,b) \oplus (0,q) = (a,b)$ for every ordered pair (a, b) and positive integer q.

Theorem 3.3: Operation $\oplus$ has an inverse (additive inverse) in $Z \times N$.

We can easily verify that for every ordered pair (a,b), $(-a,b)$ is an additive inverse.

We can easily prove that multiplication $\otimes$ is commutative and associative.

Theorem 3.4: Operation $\otimes$ has an identity (multiplicative identity) in $Z \times N$.

The reader can verify that (p,p) is a multiplicative identity, where p is a positive integer.

Theorem 3.5: All members of $Z \times N$, except for additive identity, have a multiplicative inverse.

We can easily prove that (b,a) is a multiplicative inverse of (a,b). In case of a being negative, the multiplicative inverse is $(-b, -a)$.

The reader can easily verify that multiplication $\otimes$ is distributive over addition $\oplus$.

The field of rational numbers is an ordered field. We can prove the same. We are omitting the proof.

3.6 INTERPRETATION OF RELATION AND OPERATIONS

$Z \times N$ with addition $\oplus$ and multiplication $\otimes$ is a *field* as it is a commutative ring with the existence of multiplicative inverse for all elements (except for additive identity).

Apparently, this structure does not sound like the extension of Z (the set of integers). Let us try to plot ordered pairs. We have seen that any ordered pair (a, b) is equivalent to (ap, bp), where p is a positive integer as (p, p) is a multiplicative identity. Any ordered pair can be reduced to relatively prime numbers (relatively prime numbers do not have any positive divisor other than 1).

Points corresponding to equivalent ordered pairs are on a line. If we extend the lines, they pass through the origin $(0, 0)$ though it is not a member of $Z \times N$. An interesting thing about these lines is that they have different slopes. Each of the line representing equivalent ordered pairs intersects with this line parallel to X-axis with offset 1 (equation $y=1$). All the ordered pairs $(a, 1)$ fall on line $y=1$. In the next section, we will understand this construct as the extension of integers.

3.7 MAPPING OF ORDERED PAIRS AS EXTENSION OF INTEGERS

For any ordered pair (a, b), we can find ordered pair $(c,d) \ni (a,b) = (c,d)$, and c and d do not have any positive common divisor other than 1. If a and b do not have any positive common divisor other than 1, then (a, b) itself is a required ordered pair. In the case of a and b having the greatest common positive divisor f other than 1, $a=fc$ and $b=fd$ for some integer c and positive integer d. By multiplying a by d, we get $ad=fcd$, and by multiplying b by c, we get $bc=fcd$. From this, we can say that $ad=bc$. We can say that $(a, b) = (c, d)$ as $ad=bc$.

For each ordered pair (a, b), there is an equivalent class $[(p, q)]$. We can have a partition corresponding to this equivalence relation. Let us define the partition set as $Q = \{[(p, q)] \mid p$ is an integer; q is a natural number; p and q do not have common positive integer divisor $>1\}$.

We define a function from $A = \{[(a,1)] \mid a \in Z\}$ to Z as $f([(a,1)]) = a$.

We can easily verify that the function is one-one and onto. Moreover, we can easily check that the function preserves operation as defined in the respective structure.

A subset of Q is mapped to Z, and the mapping preserves the operations. Hence, we consider Q as an extension of Z.

For each ordered pair (a, b) other than additive identity, pair $(0, 1)$ (and its equivalent) has a multiplicative inverse. So we can say that the multiplicative inverse of class $[(a, b)]$ is $[(b, a)]$, if $a>0$ else $[(-b\ -a)]$. Thus, we have an extended set of integers Z to set Q, which has a multiplicative inverse for each of the members, except for additive identity.

We will look at the multiplicative inverse of special cases for ordered pairs of the type $(b, 1)$, where b is a positive integer. For positive integers, the multiplicative inverse of $(b, 1)$ is $(1, b)$. If we multiply the equivalent of integer $(a, 1)$ by reciprocal of b, we get $(a,1) \otimes (1,b) = (a,b)$ by the definition of the multiplication operation $\otimes$. The ordered pair (a, b) represents "a divided by b." The reader can revisit the definition of operations and confirm that the equivalence and operations match the way we handle ratios of integers.

We take a special case of the ordered pair of type (a, b) and (c, b) and verify that the rational numbers with a common denominator can be added by simply adding numerators and keeping the same common denominator.

As we have seen earlier, each ordered pair can be reduced to an ordered pair of relatively prime numbers. So without the loss of generality we can restrict our working to ordered pairs with relatively prime numbers. We try to work out the optimal process for binary operations.

If we need to find $(a,b) \oplus (c,d)$ we find the least common multiple (LCM) of b and d. If the least common multiple is m then $m = b \cdot b'$ and $m = d \cdot d'$. We can easily establish that $(a,b) = (a \cdot b' + c \cdot d', m)$.

This explains the method of finding a common denominator and multiplying by the numerator of each of two numbers by (LCM divided by denominator) and adding the results with LCM as the denominator.

We look at another aspect of rational numbers. Let us consider ordered pair (a, b), where a is the positive integer. We can do integer division of a by b; then we get non-negative quotient q and remainder r such that $0 \le r < b$. This can be expressed as $a = q \cdot b + r$, where b is non-negative and $0 \le r < b$. Using the definition of equivalence and operations, we can verify that $(a,b) = (q,1) + (r,b)$.

By the definition of ordering, $0 \le (r,b) \le 1$ and r being less than b, $0 \le (r,b) < 1$. This establishes that any positive rational number can be expressed as an integer part and a fractional part between 0 (inclusive) and 1 (exclusive).

3.8 REPRESENTATION OF RATIONAL NUMBERS

One trivial representation of the rational number is *as the ratio of integers*, where the denominator is non-zero. There are some inherent difficulties in this representation. For example, adding or comparing $\dfrac{17}{25}$ and $\dfrac{43}{64}$ is not trivial.

The process to compare is to have a common denominator.

Rational numbers are represented using the decimal system having an integer part followed by the fraction separator (decimal separator), which is period "." in most of the countries and comma "," in some European countries. Position value weightages for places on the right of the separator position weightage are negative powers of 10. For example, in 23·579 the value of 5 is 5×10^{-1}, the value of 7 is 7×10^{-2}, and the value of 9 is 9×10^{-3}. This method of representing the number is popular as it is convenient to do arithmetic and comparison.

The alternate method is to put representation with a decimal point . $\dfrac{17}{25}$ is 0·68 and $\dfrac{43}{64}$ is 0·671875. If we extend them to six digits after the decimal point, they become 0·680000 and 0·671875. We have converted the number to common denominator 1000000 or 10^6. After this conversion, we can easily decide which number is smaller. In common arithmetical problems, operations are to be done several times, and every time converting all operands to a common denominator is not practical. Can we have a common denominator for all numbers? Apparently, it appears that we cannot find such a common denominator. If the rational number is the ratio of a relatively prime (no common divisor) numerator and a denominator, then the number can be converted only to the denominator, which is the multiple of the original number's denominator. At the same time, handling different denominators is difficult. For practical calculations, we agree to use an approximate value within an acceptable error limit. A simple example of using approximate values is taking the value of π as $\dfrac{22}{7}$ or 3·14 and for $\sqrt{2}$ (the length of hypotenuse for the right triangle with each side forming the right angle of 1 unit) using values 1·4 or 1·41 or 1·4142. Depending on the accuracy requirement, we select the denominator as 10^k. If the rational number is converted properly to the ratio of integers with the denominator 10^k, then the maximum difference between the original number and the number with 10^k as the denominator (error) is $\dfrac{1}{10^k}$. The algorithmic process is to multiply the original number by 10^k and take the integer part. We take an example of $\dfrac{22}{7}$.

Let us try to represent with the denominator as 10^4. The number with the denominator common for 7 and 10^4 is 7×10^4. We can represent the number with this denominator and then separate the required denominator and remaining parts as one term. The remaining term is divided into an integer part and fractional part such that the fractional part is non-negative and less than 1. Details for the selected number are given here.

$$\frac{22}{7} = \frac{22 \times 10^4}{7 \times 10^4} = \frac{22 \times 10^4}{7} \times \frac{1}{10^4} = \frac{220000}{7} \times \frac{1}{10^4}$$

$$= \frac{31428 \times 7 + 4}{7} \times \frac{1}{10^4} = \left(31428 + \frac{4}{7} \right) \times \frac{1}{10^4} = \frac{31428}{10^4} + \frac{4}{7} \times \frac{1}{10^4}.$$

We take the first part as an approximate value of the original number and discard the error term. We can clearly see (and prove) that the discarded term is less than $\dfrac{1}{10^4}$. The approximate value taken here is

$\dfrac{31428}{10^4} = \dfrac{30000 + 1428}{10000} = 3 + \dfrac{1428}{10000} = 3.1428$. The error term (difference) is

$\dfrac{4}{7} \times \dfrac{1}{10^4}$, which is less than $\dfrac{1}{10^4}$.

We can have the process as incremental. We can start with a common denominator 1. In that case, $\dfrac{22}{7} = 3 + \dfrac{1}{7}$. We take the integer part as 3 and error term as $\dfrac{1}{7}$. The error term is less than 1. If we need to improve the result, we convert it to denominator $10^1 = 10$. The error term with this denominator becomes $\dfrac{1}{7} \times \dfrac{10}{10} = \dfrac{10}{7} \times \dfrac{1}{10} = \left(1 + \dfrac{3}{7}\right) \times \dfrac{1}{10} = \dfrac{1}{10} + \dfrac{3}{7} \times \dfrac{1}{10} = 0.1 + \dfrac{3}{7} \times \dfrac{1}{10}$. This results in the error term becoming $\dfrac{3}{7} \times \dfrac{1}{10}$. From this we take significant part 0·1, and the new error term now becomes $\dfrac{3}{7} \times \dfrac{1}{10}$. This makes the result 3·1, and the error term is less than 10^{-1}. We can repeat converting the error term to the next negative power of 10. Thus, we can do a stepwise improvement in the result.

The method of representation with the decimal point is a practical solution. In this case, the numbers are written as 0·68 and 0·671875, respectively, for $\dfrac{17}{25}$ and $\dfrac{43}{64}$. Extra trailing zeros can be added, or the number of digits can be reduced to a desired level (with possible rounding).

For addition and subtraction, numbers with a fixed number of digits after the decimal point (same for both the operands) can be done like integers, and the decimal point can be inserted at the result at an appropriate place. For example, 1·5 and 2·75 can be added by first converting them to the same number of digits after the decimal place. The numbers get converted to 1·50 and 2·75. Ignoring the decimal point, they become 150 and 275. Integer addition of these numbers gives 425, which after inserting back the decimal point becomes 4·25. For multiplication, numbers are multiplied as if they do not have a decimal point and the decimal point is added at an appropriate place (the sum of the number of digits after the decimal point in both numbers). To multiply 1·5 and 2·75 numbers without the decimal point, 15 and 275 are multiplied to get the result 4125, which becomes 4·125 after inserting the decimal point at three (1+2) places from the rightmost position. Proofs of these processes are left to the reader.

The *limitation of converting the number* to 10^k as the denominator is that if the denominator of the given number has any prime factor other than 2 and 5, the process never ends. For example, a decimal representation of $\frac{1}{3}$ is never ending. It has got a never-ending sequence of digit 3. A decimal representation of $\frac{1}{7}$ is a never-ending repeating group of digits, 142857. An interesting thing about the decimal representation of rational numbers is that leaving some initial digits, it has got infinitely many times *repeating groups of digits*. The numbers for which the decimal representation ends after finite positions, we can say that it has got infinitely many times repeating 0, after the original last position having a non-zero digit.

An interesting question here is, "Is 0·999 … approximately 1 or exactly 1?" Some people believe that it is approximately 1, but the fact is that it is exactly 1. We can say that 0·999 … is infinite sum $\sum_{i=1}^{\infty} 9 \times 10^{-i}$. The limit, if exists, is the value of the infinite sum and not the approximate sum. The partial sum is the approximate value of the limit. If we consider 0·9999, it is approximately 1 (with a tolerance of 0·0001), but if we consider infinitely many times digit 9, then the value is 1.

Coming back to the decimal representation, there is a loss of precision as we cannot handle infinitely many digits. An advantage of the decimal representation is that we can do the arithmetic very easily. A simple example of the loss of precision is $\frac{1}{3} + \frac{1}{3} + \frac{1}{3}$ does not equal 1, if we use decimal representation. After any series of calculations, if the final value has an acceptable error (difference), then it is accepted in real-life problems. Hence, the decimal representation is very popular in calculations related to real-life problems.

Storage space required to store a number in binary in computers is that 10 bits can store unsigned integers (non-negative) up to 1023. Taking this range as approximately 10^3, 10 bits are required to store number up to 10^3. This can be extended to 20 bits for numbers up to 10^6 and 30 bits for numbers up to 10^9. For integer calculations, 32 bits are good enough to handle signed integers. For larger integers 64 bits can be used. Integers using 32 bits can handle signed integers up to the absolute value 2×10^9, whereas 64 bits can handle signed numbers up to the absolute value 8×10^{18}. These ranges are fairly good for routine integer calculation. However, for arithmetic with an assumed decimal point, the range reduces drastically. Another aspect of very large and small numbers is that rather than handling all digits of the number, significant digits are kept in mind. For example, number 125,436,189 is referred to as 125 million or 125 million 436 thousand or 125 and a half million. If the number given here is, say, the profit of a company for a year, then while discussing the company performance, the profit is considered as 125 million (though books of accounts have the amount up to two decimal places). Another example is that Avogadro's number is taken as $6.02214076 \times 10^{23}$. It is not practical to speak or write number with all the digits listed. This kind of representation is popularly known as the scientific format. The number is divided into two parts. The first part is a number possibly with a decimal point, and the second part is the power of 10. There is the standardized form (called normalized form) in which the number part has no digit before the decimal point and the first digit after the decimal point is non-zero. Avogadro's number in this structure can be written as the first part having (decimal point implied) 602214076 and the power part as 10^{24}. The first part having digits is called mantissa, and the second part (only the power value) is called an exponent. For Avogadro's number mantissa is 602214076 and the exponent is 24. If we take another example, number $25\dfrac{43}{64}$ written with a decimal point is 25.671875, which in normalized form is $.25671875 \times 10^2$. The number in mantissa-exponent form has mantissa 27671875 and exponent 2. Any of exponent and mantissa can have a negative sign. An advantage of this system compared to an assumed decimal point (also known as fixed decimal point) is that with defined mantissa length and exponent length, relatively large range can be covered. For example, eight-digit mantissa and two-digit exponent (both signed) can cover numbers up to 10^{99} with eight significant digits. From small absolute value point of view (other than 0), it can store 10^{-99}. In human readable form, large numbers are represented in mantissa-exponent format, known as scientific notation with one difference. Mantissa has one non-zero digit before the decimal point and the number has an explicit fraction separator character "." in most of the countries, whereas "," in some European countries it is followed by literal character E and then the exponent value. Number $25\dfrac{43}{64}$ in the *scientific notation* is $2 \cdot 567875E1$ or $+2 \cdot 567875E+1$ (if we fully qualify both mantissa and exponent with sign).

We have selected 10^k as the denominator for some non-negative integer value of k. We can select any positive integer B in place of 10. In that case, we should write all terms in the B base system. For example, for number $\dfrac{43}{64}$, we take 16^2 as an additional divisor.

Here, the stepwise process is more convenient as in every step we get one hexadecimal digit.

If we take number $\left(25\dfrac{43}{64}\right)_{10}$ (decimal system), then the number in hexa-decimal system is $(19 \cdot AC)_{16}$. We can represent the same as $0 \cdot 19AC \times B^2$, where the value of B is 16 (in decimal system). Symbol B is used to avoid confusion between decimal and hexadecimal system. The same can be represented as

0.0001 1001 1010 1100×B^8, where · is binary fraction separator and the value of B is 2. We can verify that it represents the decimal number $25\dfrac{43}{64}$.

This type of representation of numbers is called *floating-point representation*. Similar to decimal representation, values with any base is approximate with an error, which can be ignored. The actual representation in computers depends on the standard used for implementation. However, all the standards use some encoding method for exponent, mantissa, and their signs to accommodate the number in a fixed number of bits.

Programming languages like FORTRAN use data type FLOAT for floating-point numbers. Alternate keyword used for this representation is REAL. The number represented in floating-point representation is a rational expression. Possible reasoning behind using data type keyword REAL is that real-valued functions with real arguments (though they may give approximate rational value) are defined on this type of variable.

3.9 LIMITATIONS OF RATIONAL NUMBERS

The major limitation of integers, not having multiplicative inverse (except for 0), is overcome. However, equation $x^2=2$ does not have a rational solution. There are many such equations like $x^2=a$, where a is not a complete square, which do not have a rational solution.

Theorem 3.6: $x^2=2$ does not have a rational solution.

Proof: We can prove this by contradiction.

Let us assume that this equation has a rational solution and establish that it leads to a contradiction.

Let us assume that $\dfrac{p}{q}$ is the solution of the equation, where p is an integer and q is a natural number. We can safely assume that p and q have no factor in common except for 1 and −1.

Squaring the rational solution leads to p being even and subsequently q also being even. This contradicts that there is no common factor for p and q other than 1 and −1. Hence, we get a contradiction. So it is established that the equation does not have a rational solution.

3.10 SUMMARY

Some important properties of rational numbers with the operations defined here are as follows:

1. A set of rational numbers with addition and multiplication forms a field.
2. Ordering $\leq$ is the total order.
3. Any rational number representation in the decimal system has either terminating or repeating group of digits representation. If the denominator has any prime factor other than 2 and 5, then it does not have a terminating (has a repeating group) representation.
4. A rational number can be represented in any base (integer greater than 1) like integers. In addition to non-negative powers of the base as multipliers, it has negative powers of the base as multipliers on the right of fraction separator.
5. Rational number arithmetic can be done with a fixed decimal point. Addition can be done by aligning the decimal point and by adding numbers like integers, maintaining the decimal point position. For multiplication, numbers are multiplied ignoring the decimal point and inserting a decimal point at the place such that digits on the right of the decimal point are the sum of the number of digits after decimal points for individual numbers.
6. Rational numbers can be represented as floating-point numbers. This is a very efficient way of handling numbers on computers.
7. For any two different rational numbers, there is a rational number (other than the two given numbers) between the two numbers.
8. Equations like $x^2 = 2$ do not have a rational root for x.

Real Numbers

4

We have gone through the properties of rational numbers in the previous chapter. An important thing about rational numbers is that with the addition and multiplication operations, it is a field. We know that $x^2 = 2$ does not have a rational solution for x.

We go for a broader shortcoming of rational numbers rather than extending it to have the solution of the equation given above and other equations similar to it.

Definition 4.1: For a subset B of set A with the total order $\leq$ is *bounded above* if $\exists u \in A \ni \forall x \in B$, $x \leq u$ and u is an *upper bound* of set B in A.

An important point to be noted is that if u is an upper bound and if $u \leq v$ then v is an upper bound.

Example 4.1: $A = \{1, 11, 21\}$ as the subset of N with usual $\leq$ ordering is bounded above. One upper bound is 50, and another upper bound is 31.

Example 4.2: $P = \{p \mid p$ is a prime natural number$\}$ as the subset of N with usual $\leq$ ordering is not bounded above as we can always find a prime > any natural number.

Example 4.3: $B = \{x \mid x \geq 0; x^2 < 2\}$ with the usual order in Q is bounded above, and 3 is an upper bound of B in Q.

Definition 4.2: For a bounded-above subset B of set A with the partial order $\leq$ is said to have the *least upper bound* u_0 if u_0 is an upper bound and $u_0 \leq u$ if u is an upper bound of B.

Definition 4.3: For a subset B of set with the total order $\leq$ is *bounded below* if $\exists l \in A \ni \forall x \in B$, $l \leq x$ and l is a *lower bound* of set B in A.

Definition 4.4: For a bounded-below subset B of set A with the partial order $\leq$ is said to have the *greatest lower bound* l_0 if l_0 is a lower bound and $l \leq l_0$ if l is a lower bound of B.

An important point to be noted here is that if there is a least upper bound for a bounded set then it is unique. It is not necessary to have the least upper bound for all bounded sets. We are going to extend the set of rational numbers to real numbers to overcome this shortcoming of rational numbers.

4.1 LEAST UPPER BOUND PROPERTY

$B = \{x \mid x \geq 0; x^2 < 2\}$ with usual order in Q is bounded above but does not have the least upper bound. We intuitively know this fact. If we try to calculate square root using the long division method, we find value approximately 1·41421. Any rational number has square >2; then the first different digit in the decimal representation of that number has to be larger than the corresponding digit of an approximate value of the square root of 2. For example, if we start with upper bound 1·4143 then we can consider number 1·41429 (number larger than the approximate value). We can verify that 1·41429 is another upper bound, which is smaller than the earlier upper bound. We shall give formal proof now.

Theorem 4.1: Set $A = \{x \mid x \geq 0, x^2 < 2, x$ is a rational number$\}$ is bounded above and does not have the least upper bound in the set of rational numbers.

The reader can prove this.

Hint: On both sides, for value u, take next value $\dfrac{2u+2}{u+2}$.

We have found a "hole" in the set of rational numbers. In fact, there are a lot more "holes" than rational numbers. By the end of this chapter, we shall establish that the set of real numbers ("holes" plugged-in) is non-countable. Hence, additional elements in the set of real numbers are "many more" (uncountbly infinitely many elements) than rational numbers (countably infinite).

Before going for a structure, which addresses the "holes," we introduce the concept of the structure on the basis of rational numbers, which represents the set of rational numbers itself. This may sound redundant, but it forms the background for Dedekind cuts. If we consider the set of rational numbers $\{x \mid x^2 < 2\}$ then our focus is for members, where the value of x^2 is near 2. For each member of the set, there is a larger member in the set and for every member smaller rational numbers are in the set (if we consider only positive values). We define the set of rational numbers to represent each rational number with similar properties.

4.2 RATIONAL CUTS

Definition 4.5: A set $C_q = \{x \mid x < q, x \in Q\}$ is a *rational cut* corresponding to rational number q.

Important properties of rational cut are as follows:

1. Each cut is a proper subset of Q
2. $r \in C_q \wedge s \in Q \wedge s < r \Rightarrow s \in C_q$
3. $r \in C_q \Rightarrow \exists s \in C_q \ni r < s$

We can visualize this as "cutting" the line with rational numbers at a rational number and keeping only the left-side part (excluding the cutting point) and discarding the right-side part (including the cutting point).

Now we define binary operations on rational cuts.

Definition 4.6: *Addition* $\oplus$ of rational cuts C_p and C_q is defined as $C_p \oplus C_q = C_{p+q}$.

Definition 4.7: *Multiplication* $\otimes$ of rational cuts C_p and C_q is defined as $C_p \otimes C_q = C_{p \times q}$.

Now, we define Dedekind cuts, which use the basic guiding principles from rational cuts. We can represent rational numbers on a line. For rational cuts, we have taken the point representing the rational number as the "cutting point." In a cut, all the rational numbers on the left side of the "cutting point" are in the cut set, and for every point in a cut there is a point for the larger value (on the right of the line) in the cut set. We extend the scope to select the arbitrary "cutting point."

4.3 DEDEKIND CUTS

We construct the set of real numbers by extending the set of rational numbers.

Definition 4.8: A subset C of Q (set of rational numbers) is a *cut* if it satisfies the following properties:

1. C is the proper subset of Q
2. $p \in C \wedge q \in Q \wedge q < p \Rightarrow q \in C$
3. $p \in C \Rightarrow \exists r \in C \ni p < r$

Explanation: If a rational number is in a cut then all smaller rational numbers are in the cut by the second property. Moreover, for every member of a cut, there is a larger member in it.

Example 4.4: For a rational number a, set $A = \{x \mid x < a, x \in Q\}$ is a cut.

The reader can easily verify that the set A is a cut.

Theorem 4.2: For a cut C, if rational number $a \notin C$ then for rational number $b > a \Rightarrow b \notin C$.

Proof: We prove this by contradiction.

It is given that $b > a$. This is equivalent to $a < b$. If $b \in C$ then by the second property of the cut, $a \in C$, which is a contradiction as it is given that $a \notin C$. Hence $b \in C$ is not possible. So $b \notin c$.

Theorem 4.3: For a cut C, $a \notin C \Rightarrow a > b \forall b \in C$.

Proof: We prove this by contradiction.

It is given that $a \notin C$ and $b \in C$. Let us assume that $a \leq b$.

If $a = b$ then there is a contradiction with given $a \notin C$ and $b \in C$. The same rational number cannot be in the set and not in the set.

If $a < b$ then since $b \in C$, by the second property of the cut, $a \in C$. This contradicts with the given fact that $a \notin C$. Again we get a contradiction.

Our assumption of $a \leq b$ is not acceptable. So $a > b$ is acceptable. Hence for arbitrary $b \in C$ and $a \notin C \Rightarrow a > b$.

Theorem 4.4: If A and B are cuts then either $A \subseteq B$ or $B \subseteq A$.

We leave the proof to the reader.

4.4 ORDERING IN CUTS

Definition 4.9: On set R of all cuts, *ordering* $\leq$ is defined as $A \leq B$ if and only if $A \subseteq B$.

"Being subset" is a partial ordering. In general, partial ordering defined by "being subset" is not the total order. However, according to Theorem 4.4, for any two cuts, one of them is the subset of another. Hence, the ordering is the total ordering.

Definition 4.10: Cut $C_0 = \{x \mid x \in Q, x < 0\}$ is called *Zero* cut.

One can easily prove that the set defined as zero cut is a cut.

Definition 4.11: A cut C is a *negative cut* if $C < C_0$, and it is a *positive cut* if $C > C_0$.

This definition is valid as the ordering is the total order in the set of cuts. From Definition 4.11, it is clear that any cut is either negative, zero, or positive cut as $\leq$ is the total order.

4.5 BINARY OPERATIONS IN CUTS

Definition 4.12: Binary operations *addition* $\oplus$ is defined as follows:

$C \oplus D = \{x + y \mid x \in C; y \in D\}$.

The reader can easily verify that the result set expression is a cut.

The reader can prove that the addition of cuts is commutative and associative.

Theorem 4.5: Zero cut $C_0 = \{x \mid x < 0; x \in Q\}$ is an additive identity.

We leave the proof to the reader.

Theorem 4.6: For every cut C, $\exists$ cut $D \ni C \oplus D = C_0$, where C_0 is zero cut.

We leave the proof to the reader.

Definition 4.13: For cuts $C > C_0$ and $D > C_0$, *multiplication* $\otimes$ is defined as $C \otimes D = \{p \mid p \leq r \times s; r, s > 0; r \in S; s \in D\}$.

We can easily prove that the set defined here is a cut.

So far, we have defined multiplication only for positive cuts. We extend the definition for all cuts.

Definition 4.14: *Multiplication* of cuts C and D, denoted by $C \otimes D$, is defined as follows:

$$C \otimes D = \begin{cases} \{p \mid p \leq r \times s; r, s > 0; r \in C; s \in D\} & \text{if } C > C_0 \text{ and } D > C_0 \\ C_0 & \text{if } C = C_0 \text{ or } D = C_0 \\ -((-C) \otimes D) & \text{if } C < C_0 \text{ and } D > C_0 \\ -(C \otimes (-D)) & \text{if } C > C_0 \text{ and } D < C_0 \\ (-C) \otimes (-D) & \text{if } C < C_0 \text{ and } D < C_0 \end{cases}$$

We have reproduced the definition for positive cuts (cuts > additive identity C_0).

For other cuts, we have defined multiplication using the existence of additive inverse (denoted as $-X$ for the additive inverse of X). We can verify that all the expressions used in the definition are well-defined.

Distributive low (multiplication over addition) holds good. We are omitting the proof of the same.

All the requirements for being field are satisfied by the set of cuts and operations (addition and multiplication) defined over them.

Moreover, we can prove that the field of cuts is an ordered field. We do not give proof here as it is clear from the definition.

4.6 LEAST UPPER BOUND PROPERTY

Theorem 4.7: Set A of cuts bounded above has the greatest upper bound (in the set of all cuts).

Proof: It is given that A is a set of cuts, and it is bounded above. Let us define set $C = \{p \mid \exists D \in A \ni p \in D\}$. As A is bounded above, $\exists$ a cut $B \ni D \leq B \forall D \in A$. Any rational number not in B cannot be a member of any cut in the collection of cuts A. Such a rational number is not a member of set C. Hence, C is not Q itself. Moreover, it is obvious that set C is not empty. Hence, C is a proper subset of Q.

If $p \in C$ and $q < p$ then $\exists D \in A \ni p \in D$. If $q < p$ then $q \in D$. Hence, by definition of cut C, $q \in C$.

Similarly, for $p \in C$, $\exists D \in A \ni p \in D$. As $p \in D$, $\exists q > p \ni q \in D$ as D is a cut. By definition of C, $q \in C$. We have established that for every member of C we can always find a member in it greater than the given member.

The set C is a cut.

Take arbitrary member D of A. If $p \in D$ then by definition of C, $p \in D$. Hence, $D \subseteq C$. By definition of the order of cuts, $D \leq C$. As this is true for all members of A, C is an upper bound of A.

Let U be an upper bound of A. Let us take arbitrary member p of cut C. By the definition of cut C, $p \in D$ for some cut D. U being an upper bound, $D \leq U$. By the definition of the ordering of cuts, $D \subseteq U$. Hence, $p \in D$. Thus, every member of C is a member of U. So $C \subseteq U$ establishes that $C \leq U$.

We have proved that C is an upper bound, and if U is an upper bound then $C \leq U$. Hence, C is the least upper bound.

We have established that every bounded set of cuts has the least upper bound in the set of all cuts. Any bounded sets like $\{x \mid x \in Q; x > 0; x^2 < 2\}$ are bounded above in Q, but it does not have the least upper bound in Q. However, it has the least upper bound in the set of cuts.

4.7 SET OF CUTS AS EXTENSION OF RATIONAL NUMBERS

For each rational number r, we can define set $C_r = \{p \mid p < r, p \in Q\}$. We can easily verify that set C_r is cut for every rational number r. Moreover, $C_r \oplus C_s = C_{r+s}$ and $C_r \otimes C_s = C_{r \times s}$. In addition, $C_r \leq C_s \Leftrightarrow r \leq s$. Thus, we can say that cuts are the extension of rational numbers.

Cuts corresponding to rational numbers are written as the rational number used as bound in the definition of cut. Other cuts can be written by the characteristics of the definition of cut. For example cut defined as $\{p \mid p < 0 \lor p^2 < 2; \, p \in Q\}$ is expressed as $\sqrt{2}$. Those who are having an interest in the convergence of sequences will find that every bounded monotonous sequence (of rational numbers) converges to a real number. This can be related to the cut, and the convergence point can be equated to a cut. The cut can be represented by the convergence point of the sequence.

4.8 CARDINALITY OF SET OF REAL NUMBERS

We have an extended set of rational numbers to the set of all cuts, which has the least upper bound property. Intuitively, we can say that the set of cuts has more elements compared to the set of rational numbers. The natural curious question is, How many more elements are there? It is obvious that there are infinitely many more members. For example, $n\sqrt{2}$ for all natural numbers n can be represented by cuts, but they are not rational numbers. This is just one example. Square roots of all not-complete square integers are not rational. The question here is, Are the additional members countable? We shall explore the cardinality of the set of real numbers.

We have seen the decimal representation of rational numbers. We can easily prove that all the rational numbers can be expressed in a system with a decimal point, which either terminates or has a repeating group of digits. We can have a one-one and onto function from the open interval (0,1) to the set of real numbers. So, for the cardinality of the set of real numbers, we shall study cardinality of open interval (0,1).

Let us assume that this set is countable. Any countable set can be represented as the sequence of members of the set. Let us list the real numbers in interval (0,1) using their decimal representation as a sequence. Terminating representation has alternate non-terminating repeating representation. For example, we can represent $\dfrac{1}{2}$ as 0·5 or 0·49999 …. We select 0·49999... for our purpose. We can select any one of them. We form a number by changing i^{th} digit for i^{th} term of the sequence. If the digit to be changed is 5 then change it to 7, and if it is other than 5, change it to 5. To understand the process, we shall do the process for the first three numbers. If digits to be considered are only three (each number will have more digits), we list only the first three digits. Let us say these terms are 0·512, 0·125, and 0·119. In the number to be formed, the first digit should be different from the first digit of the first term. By the rule, since the first digit of the first number is 5, we pick 7 for the first position. The second digit of the second term is 2, so our selection is 5. Similarly the third digit is 5. The number formed is 0.755, which is different from the first three terms. From the rules it is obvious that the number formed up to k positions is different from first k terms. So the number formed by this method is different from all the listed terms. If the number formed by this process is appearing in the list at say m^{th} position then the number formed by this construction and the number have to be same. But, as per the construction rule, for m^{th} position, the mth digit of the constructed number and the same position digit in the term in the list have to be different. Hence, it cannot be equal to the term at the mth position. So we have found a number which is not listed. Moreover, it is not 0.9999… as 9 does not appear in the constructed number as a digit at any position. Thus, the constructed number is not 1. Thus any sequence of numbers between 0 and 1 cannot cover all numbers in the interval. Thus, interval (0,1) is not countable.

We have found that we have added uncountably infinite number of members in the set of rational numbers to get the set of real numbers. Had the additional members (other than rational numbers) been countable, the set of real numbers would have been countable as the union of countable sets is countable.

The members of the set of real numbers, which are not rational numbers, are called irrational numbers. The real numbers, which is zero of a polynomial (Solution of polynomial $P(x) = 0$) with rational coefficients, are called *algebraic numbers*. Other real numbers (obviously they are irrational numbers) are called *transcendental numbers*. Numbers like π – ratio of length of circumference to diameter of circle and $e = \dfrac{1}{0!} + \dfrac{1}{1!} + \dfrac{1}{2!} + \dfrac{1}{3!} + \cdots$ – are transcendental numbers. An interesting point to be noted is that the set of algebraic numbers is countable.

Our greed for an extension has brought us to real numbers from natural numbers. Is this the end? Though the set of real numbers sounds adequate for many of requirements, it does not have the solution for equation $x^2 + 1 = 0$. By the definition of multiplication, the square of any real numbers is positive. Hence, the equation given here does not have any real solution. We shall extend the system to complex numbers in the next chapter.

4.9 LIMITATIONS OF REAL NUMBERS

We have extended the set of rational numbers to have the least upper bound property. This makes it a complete field with the total ordering. Even after such big achievement, equations like $x^2 + 1 = 0$ do not have the solution in it. We will be extending real numbers to complex numbers in the next chapter.

4.10 SUMMARY

Set of rational numbers does not have the least upper bound property. By the introduction of cuts, we have extended the set of rational numbers to real numbers. Any set of real numbers bounded above has the least upper bound. This makes the set of real numbers a complete set. The set of real numbers, with binary operations of addition and multiplication and with ordering defined in it, makes it a totally ordered field. The set of real numbers is uncountable. The union of two countable sets gives a countable set. However, the union set of rational numbers and irrational numbers is uncountable. This establishes that the set of irrational numbers is an uncountably infinite set. In the informal language, we can say that there are "many many more" irrationals compared to rationals (pun intended!).

Complex Numbers

5

We have constructed the set of real numbers from rational numbers using Dedekind cuts. Sets of rational numbers and real numbers with addition and multiplication defined on them are ordered fields. An interesting thing about both the fields is that between any two distinct members of the set, we can always find a new number in the respective field, that is, between the two numbers. The middle number is different from both the given numbers. Rational numbers have a unique solution for single variable linear equations ($ax + b = 0$; $a \neq 0$; $a, b \in Z$) for unknown x. Some of the polynomials with rational coefficients have zeroes but not all. Equations like $x^2 = 4$ have rational solutions, whereas equations like $x^2 = 2$ do not have a rational solution. By the introduction of irrational numbers (numbers for gaps in rational numbers!), we could get the solution of $x^2 = 2$ in a set of real numbers. Even after inclusion of irrational numbers, simple polynomial equations like $x^2 + a = 0$, where a is a positive rational number, do not have a solution. In order to obtain the solution for such problems, we are going to introduce complex numbers.

5.1 COMPLEX NUMBERS AS ORDERED PAIRS OF REAL NUMBERS

We define a set of complex numbers as $C = \{(a,b) \mid a, b \in R\}$. Ordered pairs (a,b) and (c,d) are equal if and only if $a = c$ and $b = d$. This being a usual equality condition, each ordered pair is unique.

5.2 BINARY OPERATIONS IN COMPLEX NUMBERS

Definition 5.1: *Addition* $\oplus$ is defined in complex numbers as $(a,b) \oplus (c,d) = (a+c, b+d)$.

71

It is clear that the result of the addition is an ordered pair of real numbers. Hence, it is well-defined, as the value of the operation is in the set on which it is defined.

Definition 5.2: *Multiplication* $\otimes$ is defined in complex numbers as $(a,b) \otimes (c,d) = (ac - bd, ad + bc)$.

The result of multiplication being an ordered pair of real numbers, the operation is well defined as the result of operation is in the set on which it is defined.

By using the definition of multiplication of complex numbers and properties of operations on real numbers we can easily prove that addition is commutative and associative.

We can easily prove that $(0,0)$ is the additive identity and $(-a, -b)$ is the additive inverse of (a, b).

We can easily prove that multiplication is commutative and associative By using the definition of multiplication of complex numbers and properties of operations on real numbers, we can easily prove that multiplication is commutative and associative.

Theorem 5.1: Multiplicative identity exists for $\otimes$ on $R \times R$.

We can easily prove that is $(1,0)$ is a multiplicative identity.

Theorem 5.2: Multiplicative inverse exists for all members of $R \times R$ except for additive identity $(0,0)$.

We can easily prove that $\left(\dfrac{a}{a^2 + b^2}, \dfrac{-b}{a^2 + b^2} \right)$ is a multiplicative inverse of

(a, b). As (a, b) is not $(0,0)$, $a^2 + b^2$ is not equal to zero. Hence, the multiplicative inverse exists for all complex numbers except $(0,0)$.

Theorem 5.3: Multiplication is distributive over addition in $R \times R$.

Readers can easily prove this using the definition of operations and properties of real numbers.

5.3 INTRODUCTION OF IMAGINARY NUMBERS

We have defined complex numbers as ordered pairs of real numbers. Let us understand the reasoning of the definition of operations. If we take ordered pairs $(a,0)$ and $(b,0)$, $(a,0) \oplus (b,0) = (a+b,0)$ and $(a,0) \otimes (b,0) = (ab,0)$. This sounds similar to real numbers. If we take ordered pairs $(0,a)$ and $(0,b)$ then $(0,a) \oplus (0,b) = (0,a+b)$ but $(0,a) \otimes (0,b) = (-ab,0)$. So, $(a,0)$ and $(b,0)$ behave like real numbers a and b for both addition and multiplication. However, $(0,a)$ and $(0,b)$ behave like real numbers a and b for addition but not for multiplication. If we define a partial function f from $R \times R$ to R as $f(a,0) = a$. The function gets defined from subset $\{(x,0) \mid x \in R\}$ of $R \times R$. The function is one-one and onto. Moreover, it preserves the operation. Hence, we can consider $R \times R$ as an extension of a set of real numbers. If we revisit the set $((0,x) \mid x \in R)$ then the product of such two elements gives a number equated to a real number (ordered pair of type $(a,0)$). An interesting thing is that the result of the product is the same as that of multiplication of real numbers except for the reversal of sign and position in ordered pair. It reverses the sign. If we take a special case, $(0,1) \otimes (0,1) = (-1,0)$ If we consider mapping of ordered pairs to real numbers, the square of $(0,1)$ is -1. We use the symbol i for the square root of -1, which can be written as $\sqrt{-1}$.

5.4 REPRESENTATION OF COMPLEX NUMBERS

Alternate representation of ordered pair (a,b) is $a + ib$, where $a, b \in R$. This representation of complex numbers is called the Cartesian form.

Operations can be represented as $(a+ib) + (c+id) = a+c+i(b+d)$ and $(a+ib) \times (c+id) = ac - bd + i(ad+bc)$. For the complex number $a+ib$, a is called the real part and b is called the imaginary part of the number. We can use single character symbols for complex numbers. A popular symbol for complex number is z. If z is $a+ib$ then a is the real part of z and is written as Re z and the imaginary part b of z is written as Im z.

The ordered pairs can be plotted on the XY plane with the real part on the X-axis and the imaginary part on the Y-axis.

The complex number $z = a + ib$ plotted on the XY plane has a as the x coordinate and b as the y coordinate. If we draw a line segment between the origin and the point representing z, the length of the line segment is a positive value of $\sqrt{a^2 + b^2}$, represented by $|z|$. If we take one ray as the X-axis and another from the origin to the point representing z, for all numbers except for the complex number 0, then an angle is formed between two rays (from the X-axis to the ray joining the origin and the point representing the complex number). These two rays make the angle θ (signed, taking a positive value for anticlockwise rotation and usually in theoretical mathematics the angle is specified in radians). The angle for any point is between 0 and 2π; 0 is included and 2π is excluded. We can take any interval with a width of 2π and with one end-point included. Using simple trigonometry, we can easily derive that $a = |z| \cos\theta$ and $b = |z| \sin\theta$, where $|z| = \sqrt{a^2 + b^2}$. Representing the point on the plane using this approach is called polar representation. It is denoted by (r, θ), where r is the radius of the circle having origin as the center and passing through the point corresponding to the complex number and θ is the angle from the X+ve ray to the ray connecting the origin and the point of the complex number. We can say that if the usual co-ordinates (called Cartesian coordinates) are (x,y) and polar co-ordinates are (r, θ) then $x = r\cos\theta$ and $y = r\sin\theta$, where r is $\sqrt{x^2 + y^2}$ In this representation, r is called the modulus and θ is called the argument. For the given polar co-ordinates (r, θ) we can easily find (x,y) co-ordinates. Conversely, for (x,y) coordinates, $r = \sqrt{x^2 + y^2}$ and value of θ is split into three cases. If x is 0 then θ is $\dfrac{\pi}{2}$ if y is positive; else $\dfrac{3\pi}{2}$. For a positive value of y it is $\tan^{-1}\dfrac{y}{x}$ and for the negative value of y, it is $\pi + \tan^{-1}\dfrac{y}{x}$, $\tan^{-1}$ taken between 0 and π. Depending on the target interval of θ, its value is adjusted by difference of 2π.

Multiplication of complex numbers has an interesting behavior in polar representations.

Let us take two complex numbers $z_1 = (r_1, \theta_1)$ and $z_2 = (r_2, \theta_2)$. We do not have a definition for the multiplication using polar systems. Let us convert the numbers to the Cartesian form. In the target form, $z_1 = r_1 \cos\theta_1 + ir_1 \cos\theta_1$ and $z_2 = r_2 \cos\theta_2 + ir_2 \sin\theta_2$, the result of the multiplication is $r_1 r_2 \cos\theta_1 \cos\theta_2 - r_1 r_2 \sin\theta_1 \sin\theta_2 + i(r_1 r_2 \cos\theta_1 \sin\theta_2 + r_1 r_2 \cos\theta_2 \sin\theta_1)$. Using the formulae for trigonometric functions of addition, the result gets simplified to $r_1 r_2 \cos(\theta_1 + \theta_2) + ir_1 r_2 \sin(\theta_1 + \theta_2)$, which can be represented in the polar system as $(r_1 r_2, \theta_1 + \theta_2)$. In the polar system $(r_1, \theta_1) \times (r_2, \theta_2) = (r_1 r_2, \theta_1 + \theta_2)$. Multiplication of complex numbers can be done conveniently in the Cartesian form $a + ib$ and the polar form (r, θ).

We know that 1 as a complex number can be represented in the polar form as $(1,0)$. For the arbitrary non-zero complex number (r, θ), we try to find the multiplicative inverse of it. Let us say the multiplicative inverse is (s, β). By multiplication rule of complex numbers in polar form, $(r,\theta) \times (s,\beta) = (rs, \theta + \beta)$. Comparing this with $(1,0)$, we get equations $rs = 1$ and $\theta + \beta = 0$ (or $\theta + \beta = 2k\pi$, where k is any integer, to be precise), giving solution $s = \dfrac{1}{r}$ and $\beta = -\theta$. Thus the multiplicative inverse of (r, θ) is $\left(\dfrac{1}{r}, -\theta\right)$, subject to adjusting the argument by an integer multiple of 2π depending on the target interval for argument.

The polar formula for multiplication can be extended for multiplication of several complex numbers. Put in simple words, the modulus of product is the product of the moduli and the argument is the sum of arguments.

In a special case, where all numbers in the product are the same, the product becomes the nth power of the number. The formula becomes $(r, \theta)^n = (r^n, n\theta)$. This formula is much more convenient to use as binomial expansion is very lengthy compared to this. A simple step gives us the formula for the nth root. Using the formula, the modulus of the nth root is that of the given number and the argument of the nth part of an argument is that of the given number. This is correct but not exhaustive. We specify the argument from an interval of width 2π. The argument value is selected from the interval $[0, 2\pi)$ or $(-\pi, \pi]$. If a complex number is represented in polar form as (r, θ) the same can be represented as $(r, 2k\pi + \theta)$, where k is an integer. To find the nth root of (r, θ), we can take the number as $(r, 2k\pi + \theta)$. If we select the integer value of k from 0 to $n-1$, for each of the representation, we can find the nth root for each integer value of k from 0 to $(n-1)$. Thus, we have found all n distinct n^{th} roots of (r, θ) and they are for integer values of k from 0 to $n-1$. Readers can verify that all these values are distinct for integer values of k from 0 to $n-1$.

An interesting way of finding all n^{th} (for $n > 2$) complex roots of unity (*real number* 1) is by inscribing a regular polynomial with n vertices (each vertex on the circumference of the unit circle), vertices in unit circle with one vertex $(1,0)$ as $(1,0)$ is one of the n roots of 1. Each vertex of this polynomial is n^{th} root of unity.

For a given complex number $z_1 = a + ib$, can we find out if $z_2 = c + id$ ∋ $z_1 + z_2$ and $z_1 \times z_2$ are real numbers? We are interested in this for numbers other than real numbers as this property holds good trivially for real numbers. Let us take the condition on $z_1 + z_2$ first. For a number to be real, its imaginary part has to be 0. $\text{Im}(z_1 + z_2)$ is $b + d$. So, $\text{Im}(z_1 + z_2) = 0$ if and only if $b + d = 0$. Hence, the required condition is $d = -b$. Using this condition, z_2 becomes $c + i(-b)$. With these values, $z_1 z_2 = (a + ib) \times (c + i(-b)) = ac + b^2 + i(-ab + bc)$. For imaginary parts of $z_1 z_2$ to be 0, $bc = ab$ should be satisfied. In the case of $b = 0$, both the numbers are real numbers; they give the real number as a result of addition and multiplication. Our primary interest is for non-trivial cases, where $b \neq 0$. In that case, the condition gets transformed to $c = a$. Putting values in z_2, it becomes $a - ib$. For a given complex number $a + ib$, its addition and multiplication with $a - ib$ give the real number as a result. This interesting number is called its conjugate.

Definition 5.3: The *Conjugate* of complex number $z = a + ib$ is $a - ib$ and it is denoted by $\bar{z}$.

If a complex number $a + ib$ is (r, θ) then $a = r\cos\theta$ and $b = r\sin\theta$. If we take the complex number $(r, -\theta)$, the Cartesian representation of it is $c + id$ then $c = r\cos(-\theta)$ and $d = r\sin(-\theta)$. Using trigonometric identities, $c = r\cos\theta$ and $d = -r\sin\theta$. From values of a and b in terms of r and θ, $c = a$ and $d = -b$. Thus, we have proved that $(r, -\theta)$ is the conjugate of (r, θ).

If we look at a plot of complex numbers, for a complex number, its conjugate is the mirror image considering X-axis as the mirror. Readers are advised to explore properties of the conjugate in the context of operations of them and other general properties.

5.5 ORDERING IN COMPLEX NUMBERS

For a field to be totally ordered, the result of addition and multiplication of two positive (> 0) numbers must be positive. If we consider a complex number as an extension of real numbers, it should maintain the ordering of real numbers. As per the ordering of real numbers, $1 > 0$ and $-1 < 0$. Let us consider i. If i is positive then $i \times i = i^2$ should be positive. However, $i^2 = -1$. The result of the multiplication of two positive numbers becomes negative. Hence, i cannot be considered as positive. Similarly, if we assume that i is negative then $-i$ is positive by properties of the ordered field. Again $(-i)(-i) = i^2 = -1$ is negative. Hence, in both the cases, the property of the result of addition and multiplication of positive numbers being positive is not maintained. So, the field of complex numbers is not totally ordered.

5.6 CARDINALITY OF THE SET OF COMPLEX NUMBERS

One can prove that the cardinality of the interval of real numbers $(0,1)$ and the Cartesian product of intervals $(0,1) \times (0,1)$ are the same. Further, it can be proved that the cardinality of $(0,1)$ and the set of real numbers R is the same. Similarly, the cardinality of $(0,1) \times (0,1)$ and the set of complex numbers is the same. Thus, the cardinality of the set of real numbers and that of complex numbers is the same. The same is denoted by the symbol $\aleph_1$.

5.7 ALGEBRAIC NUMBERS

Any polynomial $P(x)$ of any order n with all rational coefficients has got at the most n real zeros (roots of equation $P(x)=0$). In some cases, there could be a repeated root resulting in apparently lesser no of roots. Equations like $x^2 - 2x + 1 = 0$ has one root 1, repeated twice. But some equations like $x^3 - 1 = 0$ has only one real root and the same is not repeated. Equations like $x^2 + 1 = 0$ do not have any real root.

We can find all zeroes of such polynomials in the field of complex numbers. There could be repeated roots but in all, there are as many roots as the degree of the polynomial.

The complex numbers, which are zero of some polynomial with rational co-efficients, are called algebraic numbers. A set of complex algebraic numbers is countable.

5.8 SUMMARY

The field of complex numbers can be considered as an extension of the field of real numbers. This field has the following important properties:

1. Natural numbers, integers, rational numbers, and real numbers can be represented on-line, whereas complex numbers are represented on the Cartesian plane with the real part on the X-axis and the imaginary part on the Y-axis.
2. The square root of -1 is called the imaginary number, and it cannot be obtained by adding or multiplying two real numbers. It is represented on an independent orthogonal axis compared to real numbers. A linear combination, which is formed by multiplying an imaginary number by a real number (stretching/compressing) and then adding the same to another real number, becomes a complex number.
3. The addition of complex numbers is done by adding the real part to the real part and the imaginary part to the imaginary part. Multiplication is defined as the real part having the product of both the imaginary parts (real value) subtracted from the product of both the real parts and the imaginary part as the sum of the product of the real part of the first number and the imaginary part of the second

number and the product of the imaginary part of the first number and the real part of the second number.

4. One of the popular representations of complex numbers is the Cartesian representation. In this representation, complex numbers are represented as $a + ib$. Another popular representation is the polar representation written as (r, θ), where r is the length of the line segment joining the origin $(0,0)$ representing 0 and the point representing the complex number in the XY-plane; it is called the modulus of complex numbers. θ is the angle (anticlockwise positive) from the X+ve axis ray to the ray joining the origin and the point representing the complex number. One can convert a complex number from the Cartesian form to a polar form and vice-versa.

5. The product of complex numbers in polar form can be done by a simple formula. The modulus of the product is the product of moduli of both the numbers and the argument is the sum of arguments of the numbers. This can be used in finding the nth power and n^{th} root easily. One can find all n^{th} roots of complex numbers by adding multiples of 2π in the argument of input number and dividing each value by n.

Index

Binary Coded Decimal, *see* Natural numbers, BCD system
Binary operation, 10
  associative, 11
  closure, 11
  commutative, 11
  distributive, 12
  identity, 11
  inverse of, 11
  partial, 10
Binary system, 15
  addition, 27

Cardinality, 8
Complement
  ten's complement, 44
  two's complement, 45

Dedekind cuts, 63
  addition, 65
  definition, 63
  least upper bound property, 66
  multiplication of positive cuts, 65
  ordering, 64
  positive cut, 64
  zero cut, 64

Equivalent class, 4

Field, 13
  ordered field, 13
Function, 6
  bijection, 7
  composite function, 7
  identity function, 7
  into, 7
  inverse, 8
  invertible, 7
  one-one, 7
  onto, 7
  partial function, 7

Group, 12
  Abelian group, 12
  commutative group (*see* Group, Abelian Group)
Groupoid, 12
  partial, 12

Homeomorphism, 14

Integers
  operations in ordered pairs, 36
  ordered pairs as extension of natural numbers, 42
  as ordered pairs of natural numbers, 35
  ring under addition and multiplication, 39
  ten's complement, 44
  two's complement, 45
Isomorphism, 14

Magma, *see* Groupoid
Monoid, 12

Natural numbers
  addition of, 17
  BCD addition, 28
  BCD system, 27
  binary system, 25
  first principle of mathematical induction, 20
  hexadecimal, 27
  hexadecimal system, 23
  limitations, 21
  multiplication, 18
  octal system, 27
  ordering, 19, 21
  Peano axioms, 17
  representation of, 21
  second principle of mathematical induction, 20
  set theoretic definition of, 16
  subtraction, restricted, 19
  well ordering principle, 21

Poset, *see* Set, partially ordered

Rational cut, 62
    addition, 63
    multiplication, 63
Rational numbers
    addition, 50
    with common denominator, 52
    decimal representation-repeating group of
        digits or terminating, 56
    extension of integers, 52
    field, 51
    floating point representation, 58
    limitation of converting to
        decimal, 55
    limitations of, 58
    multiplication, 50
    ordering in ordered pairs, 50
    as pairs of integers, 49
    as ratio of integers, 53
    scientific notation, 57
Real numbers
    algebraic numbers, 68
    cardinality of, 67
    limitations of, 69
    transcendental numbers, 68
Relation, 3
    antisymmetric, 5
    argument (*see* Relation, pre-image)
    co-domain of, 6
    domain of, 6
    equivalence relation, 4
    image, 6
    many-to-one, 6
    one-to-many, 6

    one-to-one, 6
    partial order, 5
    pre-image, 6
    range of, 6
    reflexive, 4
    symmetric, 4
    total order, 5
    transitive, 4
Ring, 13
    ordered ring, 13
    semiring, 13

Set, 1
    bounded above, 61
    bounded below, 61
    Cartesian product, 3
    countable, 8
    definition, 1
    empty set, 2
    finite, 8
    greatest lower bound, 61
    infinite, 8
    intersection of, 3
    least upper bound, 61
    lower bound, 61
    null set (*see* Set, empty set)
    partially ordered, 5
    partition of, 5
    power set, 3
    roster method, 2
    set builder, 2
    subset, 2
    union of, 3
    universal set, 3
    upper bound, 61

Printed in the United States
by Baker & Taylor Publisher Services